CRM Series in Mathematical Physics

Springer
New York
Berlin
Heidelberg
Barcelona
Hong Kong
London
Milan
Paris
Singapore
Tokyo

CRM Series in Mathematical Physics

Conte, The Painlevé Property: One Century Later

Jackiw, Lectures on Fluid Dynamics: A Particle Theorist's View of Supersymmetric, Non-Abelian, Noncommutative Fluid Mechanics and d-Branes

MacKenzie, Paranjape, and Zakrzewski, Solitons: Properties, Dynamics, Interactions, Applications

Saint-Aubin and Vinet, Algebraic Methods in Physics: A Symposium for the 60th Birthdays of Jiří Patera and Pavel Winternitz

Saint-Aubin and Vinet, Theoretical Physics at the End of the Twentieth Century: Lecture Notes of the CRM Summer School, Banff, Alberta

Semenoff and Vinet, Particles and Fields

van Diejen and Vinet, Calogero–Moser Sutherland Models

Roman Jackiw

Lectures on Fluid Dynamics

A Particle Theorist's View of Supersymmetric, Non-Abelian, Noncommutative Fluid Mechanics and d-Branes

Springer

Roman Jackiw
Center for Theoretical Physics
Massachusetts Institute of Technology
Cambridge, MA 02139
USA
jackiw@lns.mit.edu

Library of Congress Cataloging-in-Publication Data
Jackiw, Roman W.
Lectures on fluid dynamics : a particle theorist's view of supersymmetric, non-Abelian, Noncommutative fluid mechanics and d-branes / Roman Jackiw.
p. cm. — (The CRM series in mathematical physics)
Includes bibliographical references and index.
ISBN 0-387-95422-8 (acid-free paper)
1. Fluid dynamics. 2. Supersymmetry. 3. Fluid mechanics. 4. String models. I. Title. II. CRM series in mathematical physics.
QA911 .J33 2002
532'.05—dc21 2001059794

ISBN 0-387-95422-8 Printed on acid-free paper.

Printed in the United States of America.

9 8 7 6 5 4 3 2 1 SPIN 10863816

Typesetting: Pages created by the author using a Springer TEX macro package.

www.springer-ny.com

Springer-Verlag New York Berlin Heidelberg
A member of BertelsmannSpringer Science+Business Media GmbH

Series Preface

The Centre de recherches mathématiques (CRM) was created in 1968 by the Université de Montréal to promote research in the mathematical sciences. It is now a national institute that hosts several groups and holds special theme years, summer schools, workshops, and a post-doctoral program. The focus of its scientific activities ranges from pure to applied mathematics and includes statistics, theoretical computer science, mathematical methods in biology and life sciences, and mathematical and theoretical physics. The CRM also promotes collaboration between mathematicians and industry. It is subsidized by the Natural Sciences and Engineering Research Council of Canada, the Fonds FCAR of the Province of Québec, the Canadian Institute for Advanced Research, and has private endowments. Current activities, fellowships, and annual reports can be found on the CRM Web page at www.CRM.UMontreal.CA.

The CRM Series in Mathematical Physics includes monographs, lecture notes, and proceedings based on research pursued and on events held at the CRM.

Yvan Saint-Aubin
Montréal

Preface

This monograph is derived from a series of six lectures which were given at the Centre de recherches mathématiques (CRM) in Montréal, in March and June 2000, while the author was holder of the Aisenstadt Chair.

Précis

During the March 2000 meeting of the Workshop on Strings, Duality, and Geometry in Montréal, Canada, three lectures were delivered on topics in fluid mechanics, while the author was holder of the Aisenstadt Chair. Three more lectures were presented in June 2000, during the Montréal Workshop on Integrable Models in Condensed Matter and Nonequilibrium Physics. The following are brief descriptive remarks on the content of these lectures.

1. **Introduction.** The motivation for the research is explained.

2. **Classical Equations.** The classical theory is reviewed, but in a manner different from textbook discussions.

    2.1. *Equations of motion.* Summary of conservation and Euler equations.

    2.2. *A word on canonical formulations.* An advertisement of the method for finding the canonical structure for the above (developed with L.D. Faddeev).

    2.3. *The irrotational case.* C. Eckart's Lagrangian and a relativistic generalization for vortex-free motion.

    2.4. *Nonvanishing vorticity and the Clebsch parametrization.* In the presence of vorticity, the velocity Chern–Simons term (kinetic helicity) provides an obstruction to the construction of a Lagrangian for the motion. C.C. Lin's method overcomes this obstruction, and leads to the Clebsch parametrization for the velocity vector.

    2.5. *Some further remarks on the Clebsch parametrization.* The properties and peculiarities of this presentation for a 3-vector.

3. **Specific Models.** Nonrelativistic and relativistic fluid mechanics in spatial dimensions greater than one.

   3.1. *Galileo-invariant nonrelativistic model.* The Chaplygin gas [negative pressure, inversely proportional to density] is studied, selected solutions are presented, and unexpected symmetries are identified.

   3.2. *Lorentz-invariant relativistic model.* The scalar Born–Infeld model is found to be the relativistic generalization of the Chaplygin gas, and shares with it unexpected symmetries.

   3.3. *Some remarks on relativistic fluid mechanics.* The dynamics for isentropic relativistic fluids is given a Lagrangian formulation, and the Born–Infeld model is fitted into that framework.

4. **Common Ancestry: The Nambu–Goto Action.** Both the Chaplygin gas and the Born–Infeld model devolve from the parametrization-invariant Nambu–Goto action, when specific parametrization is made.

   4.1. *Light-cone parametrization.* Chaplygin gas is derived.

   4.2. *Cartesian parametrization.* The Born–Infeld model is derived.

   4.3. *Hodographic transformation.* Chaplygin gas is derived (again).

   4.4. *Interrelations.* The Chaplygin gas and the Born–Infeld model are related because: (1) the former is the nonrelativistic limit of the latter; and (2) both descend from the same Nambu–Goto action.

5. **Supersymmetric Generalization.** Fluid mechanics enhanced by supersymmetry.

   5.1. *Chaplygin gas with Grassmann variables.* Vorticity is parametrized by Grassmann variables, which act like Gaussian potentials of the Clebsch parametrization.

5.2. *Supersymmetry.* Supercharges, transformations generated by them, and their algebra.

5.3. *Supermembrane connection.* Supermembrane Lagrangian in three spatial dimensions.

5.4. *Hodographic transformation.* Supersymmetric Chaplygin gas in two spatial dimensions is derived.

5.5. *Light-cone parametrization.* Supersymmetric Chaplygin gas in two spatial dimensions is derived (again).

5.6. *Further consequences of the supermembrane connection.* Hidden symmetries of the supersymmetric model.

6. One-Dimensional Case. The previous models in one spatial dimension are completely integrable.

6.1. *Specific solutions for the Chaplygin gas on a line.* Some special solutions are presented; an infinite number of constants of motion is identified; Riemann coordinates are introduced and the fluid equations, as well as constants of motion, are expressed in terms of them.

6.2. *Aside on the integrability of the cubic potential in one dimension.* The one-dimensional problem with pressure $\propto$ (density)^3 possesses the SO(2,1) "Schrödinger symmetry" and the equations of motion, in Riemann form, become free.

6.3. *General solution for the Chaplygin gas on a line.* Solution obtained by linearization.

6.4. *Born–Infeld model on a line.* When formulated in terms of its Riemann coordinates, it becomes trivially equivalent to the Chaplygin gas.

6.5. *General solution of the Nambu–Goto theory for a* ($d = 1$)*-brane* (*string*) *in two spatial dimensions* (*on a plane*). The explicit string solution is transformed by a hodographic transformation to the Chaplygin gas solution, and a relation is established between this solution and the one found by linearization.

6.6. *Supersymmetric Chaplygin gas on a line.* The supersymmetric lineal Chaplygin gas, which descends from a superstring on a plane, is completely integrated, and the algebra of its conserved charges is described.

7. Toward a Non-Abelian Fluid Mechanics. Motivation for this theory is given.

7.1. *Proposal for non-Abelian fluid mechanics.* A Lagrangian is proposed; it involves a non-Abelian auxiliary field whose Chern–Simons density should be a total derivative.

7.2. *Non-Abelian Clebsch parametrization (or, casting the non-Abelian Chern–Simons density into total derivative form).* Total derivative form for the non-Abelian Chern–Simons density is found, thereby generalizing the Abelian Clebsch parametrization, which achieves a total derivative form for the Abelian density.

7.3. *Proposal for non-Abelian magnetohydrodynamics.* Our proposal, which generalizes the proposal in Section 7.1 to include a dynamical non-Abelian gauge field, reduces in the Abelian limit to conventional magnetohydrodynamics.

8. Noncommutative Fluid Mechanics. Noncommutative field theories are widely studied owing to their relation with string theory. But it is possible to present these ideas in the more mundane setting of fluid mechanics in the presence of a strong magnetic field.

8.1 *Noncommuting particles.* As an introduction to the fluid mechanical construction, we review the problem of a charged particle in a strong magnetic field and show that, after projection of the motion onto the lowest Landau level, the particle's coordinates cease to (Poisson) commute.

8.2 *Noncommuting fluids.* The particle model is generalized to a fluid model.

8.3 *Quantization.* The classical (=Poisson) noncommutativity is promoted to a quantum noncommutativity. The "star product" of noncommuting field theory is found to play a role in the description.

8.4 *Various algebras.* We summarize the interesting algebras that are encountered in our presentation.

Contents

1

Introduction

Field theory, as developed by particle physicists in the last quarter century, has enjoyed a tremendous expansion in concepts and calculational possibilities.

We learned about higher and unexpected symmetries, and discovered evidence for partial or complete integrability facilitated by these symmetries. We appreciated the relevance of topological ideas and structures, like solitons and instantons, and introduced new dynamical quantities, like the Chern–Simons terms in odd-dimensional gauge theories. We enlarged and unified numerous degrees of freedom by introducing organizing principles such as non-Abelian symmetries and supersymmetries. Indeed, application of field theory to particle physics has now been replaced by the study of fundamentally extended structures like strings and membranes, which bring with them new mathematically intricate ideas.

Thinking about research possibilities, it was decided to investigate whether the novelties that we have introduced into particle physics field theory can be used in a different, nonparticle physics, yet still in a field-theoretic context. In these Aisenstadt lectures an approach was described to fluid mechanics, which is an ancient field theory, but which can be enhanced by the ideas that we gleaned from particle physics.

As an introduction, we shall begin with a review of the classical theory. Mostly, we duplicate what can be found in textbooks, but perhaps the emphasis will be new and different. After this we shall describe how some instances of the classical theory are related to d-branes and how this relation explains some integrability properties of various models. We shall then show how the degrees of freedom can be enlarged to accommodate supersymmetry, non-Abelian structures and noncommutativity in fluid mechanics. A few problems are scattered throughout; solutions are given at the end of the text, before the References.

New work that will be described here was done in collaboration with D. Bazeia, V.P. Nair, S.-Y. Pi, and A.P. Polychronakos. Textbooks for the classical theory, which are recommended, are by Landau and Lifshitz [1] as well as by Arnold and Khesin [2].

2

Classical Equations

2.1 Equations of Motion

We begin with nonrelativistic equations that govern a matter density field $\rho(t, \boldsymbol{r})$ and a velocity field vector $\boldsymbol{v}(t, \boldsymbol{r})$, taken in any number of dimensions. The equations of motion comprise a continuity equation,

$$\frac{\partial}{\partial t}\rho(t, \boldsymbol{r}) + \boldsymbol{\nabla} \cdot \big(\rho(t, \boldsymbol{r})\boldsymbol{v}(t, \boldsymbol{r})\big) = 0, \tag{2.1}$$

which ensures matter conservation, that is, time independence, of $N = \int dr\, \rho$, and Euler's equation, which is the expression of a nonrelativistic force law

$$\frac{\partial}{\partial t}\boldsymbol{v}(t, \boldsymbol{r}) + \boldsymbol{v}(t, \boldsymbol{r}) \cdot \boldsymbol{\nabla}\boldsymbol{v}(t, \boldsymbol{r}) = \boldsymbol{f}(t, \boldsymbol{r}). \tag{2.2}$$

Here $\rho\boldsymbol{v}$ is the current $\boldsymbol{j}$ and $\boldsymbol{f}$ is the force. We shall deal with an isentropic fluid, that is, entropy is constant and does not appear in our theory. Also we ignore dissipation and take the force to be given by the pressure P: $\boldsymbol{f} = -(1/\rho)\boldsymbol{\nabla} P$. For isentropic motion P is a function only of ρ, so $\boldsymbol{f}$ can also be written as $-\boldsymbol{\nabla} V'(\rho)$:

$$\boldsymbol{f} = -\frac{1}{\rho}\boldsymbol{\nabla} P = -\boldsymbol{\nabla} V'(\rho), \tag{2.3}$$

with the dash (also known as "prime") designating the derivative with respect to argument. $V'(\rho)$ is the enthalpy, $\rho V'(\rho) - V(\rho) = P(\rho)$, and $\sqrt{P'(\rho)} \equiv s$ is the speed of sound. (Those familiar with the subject will recognize that an Eulerian rather than a Lagrangian description of a fluid is used [3].)

The dynamics summarized in (2.1) and (2.2) and the definition (2.3) may be presented as continuity equations for an energy momentum tensor. The energy density $\mathcal{E} = T^{00}$:

$$\mathcal{E} = \tfrac{1}{2}\rho v^2 + V(\rho) = T^{00}, \tag{2.4a}$$

together with the energy flux

$$T^{j0} = \rho v^j (\tfrac{1}{2} v^2 + V'), \tag{2.4b}$$

obey

$$\frac{\partial}{\partial t} T^{00} + \partial_j T^{j0} = 0. \tag{2.4c}$$

Similarly, the momentum density, which in the nonrelativistic theory coincides with the current,

$$\mathcal{P}^i = \rho v^i = T^{0i} \tag{2.5a}$$

and the stress tensor T^{ij}:

$$T^{ij} = \delta^{ij}(\rho V' - V) + \rho v^i v^j = \delta^{ij} P + \rho v^i v^j, \tag{2.5b}$$

satisfy

$$\frac{\partial}{\partial t} T^{0i} + \partial_j T^{ji} = 0. \tag{2.5c}$$

Note that $T^{0i} \neq T^{i0}$ because the theory is not Lorentz invariant, but $T^{ij} = T^{ji}$ because it is invariant against spatial rotations. [Thus $T^{\mu\nu}$ is not, properly speaking, a "tensor," but an energy-momentum "complex."]

A simplification occurs for the irrotational case when the vorticity

$$\omega_{ij} \equiv \partial_i v^j - \partial_j v^i, \tag{2.6}$$

vanishes. For then the velocity can be given in terms of a velocity potential θ,

$$\boldsymbol{v} = \boldsymbol{\nabla}\theta \tag{2.7}$$

and (2.2) can be replaced by the Bernoulli equation.

$$\frac{\partial \theta}{\partial t} + \frac{v^2}{2} = -V'(\rho). \tag{2.8}$$

The gradient of (2.8) gives (2.2), with the help of (2.3) and (2.7).

Problem 2.1. In the free Schrödinger equation for a unit-mass particle, $i\hbar\, \partial\psi/\partial t = -\frac{1}{2}\hbar^2 \nabla^2 \psi$, set $\psi = \rho^{1/2} e^{i\theta/\hbar}$, and separate real and imaginary parts. Show that the resulting equations are like those of fluid mechanics. What is the velocity? Is vorticity supported? What is the force $\boldsymbol{f}$?

Our equations can be presented in any dimensionality, but we shall mostly consider the cases of three, two, and one spatial dimensions. In the first case, the vorticity is a (pseudo)-vector

$$\boldsymbol{\omega} = \boldsymbol{\nabla} \times \boldsymbol{v}, \tag{2.9}$$

in the second, it is a (pseudo)-scalar

$$\omega = \varepsilon^{ij}\, \partial_i v^j, \tag{2.10}$$

while the last, lineal case is always simple because there is no vorticity and the velocity can always be written as the derivative (with respect to the single spatial variable) of a potential.

The dynamics of any particular system is most economically presented when a canonical/action formulation is available. To this end, we note that the above equations of motion can be obtained by (Poisson) bracketing with the Hamiltonian

$$H = \int dr \left(\tfrac{1}{2}\rho v^2 + V(\rho)\right) = \int dr\, \mathcal{E}, \tag{2.11}$$

$$\frac{\partial \rho}{\partial t} = \{H, \rho\}, \tag{2.12}$$

$$\frac{\partial \boldsymbol{v}}{\partial t} = \{H, \boldsymbol{v}\}, \tag{2.13}$$

provided the nonvanishing brackets of the fundamental $(\rho, \boldsymbol{v})$ variables are taken to be

$$\begin{aligned} \{v^i(\boldsymbol{r}), \rho(\boldsymbol{r}')\} &= \partial_i \delta(\boldsymbol{r} - \boldsymbol{r}'), \\ \{v^i(\boldsymbol{r}), v^j(\boldsymbol{r}')\} &= -\frac{\omega_{ij}(\boldsymbol{r})}{\rho(\boldsymbol{r})} \delta(\boldsymbol{r} - \boldsymbol{r}'). \end{aligned} \tag{2.14}$$

(The fields in curly brackets are at equal times, hence the time argument is suppressed.) An equivalent, more transparent version of the algebra (2.14) is satisfied by the field momentum density,

$$\mathcal{P} = \rho \boldsymbol{v}. \tag{2.15}$$

As a consequence of (2.14) we have

$$\begin{aligned} \{\mathcal{P}^i(\boldsymbol{r}), \rho(\boldsymbol{r}')\} &= \rho(\boldsymbol{r}) \partial_i \delta(\boldsymbol{r} - \boldsymbol{r}'), \\ \{\mathcal{P}^i(\boldsymbol{r}), \mathcal{P}^j(\boldsymbol{r}')\} &= \mathcal{P}^j(\boldsymbol{r}) \partial_i \delta(\boldsymbol{r} - \boldsymbol{r}') + \mathcal{P}^i(\boldsymbol{r}') \partial_j \delta(\boldsymbol{r} - \boldsymbol{r}'). \end{aligned} \tag{2.16}$$

This is the familiar algebra of momentum densities. One verifies that the Jacobi identity is satisfied [4].

Naturally one asks whether there exists a Lagrangian whose canonical variables lead to the Poisson brackets (2.14) or (2.16) and to the Hamiltonian (2.11). In more mathematical language, we seek a canonical 1-form and a symplectic 2-form that lead to the algebra (2.14) or (2.16).

Problem 2.2. Second-quantized Schrödinger fields satisfy equal-time commutation (anticommutation) relations, when describing bosons (fermions): $\lceil \psi(\boldsymbol{r}), \psi^*(\boldsymbol{r}')\rfloor_\pm = \delta(\boldsymbol{r} - \boldsymbol{r}')$. Show that the algebra (2.16) is reproduced (apart from factors of $i\hbar$) when $\rho = \psi^*\psi$, $\boldsymbol{\mathcal{P}} = \operatorname{Im} \hbar\psi^* \boldsymbol{\nabla}\psi$. Since in the nonrelativistic theory $\boldsymbol{\mathcal{P}} = \boldsymbol{j}$, find $\boldsymbol{j}$ in terms of ρ and $\boldsymbol{v}$, with $\boldsymbol{v}$ as determined in Problem 2.1.

2.2 A Word on Canonical Formulations

We shall now describe an approach to canonical formulations of dynamics, publicized by Faddeev and Jackiw [5], which circumvents and simplifies the more elaborate approach of Dirac.

We begin with a Lagrangian that is first-order in time. This entails no loss of generality because all second-order Lagrangians can be converted to first-order by the familiar Legendre transformation, which produces a Hamiltonian: $H(p, q) = p\dot{q} - L(\dot{q}, q)$, where $p \equiv \partial L / \partial \dot{q}$ (the over-dot designates the time derivative). The equations of motion obtained by taking the Euler–Lagrange derivative with respect to p and q of the Lagrangian $L(\dot{p}, p; \dot{q}, q) \equiv p\dot{q} - H(p, q)$ coincide with the "usual" equations of motion obtained by taking the q Euler–Lagrange derivative of $L(\dot{q}, q)$. [In fact, $L(\dot{p}, p; \dot{q}, q)$ does not depend on $\dot{p}$.] Moreover, some Lagrangians possess only a first-order formulation (e.g., Lagrangians for the Schrödinger or Dirac fields; also the Klein–Gordon Lagrangian in light-cone coordinates is first order in the light-cone "time" derivative).

Denoting all variables by the generic symbol ξ^i, the most general first-order Lagrangian is

$$L = a_i(\xi)\dot{\xi}^i - H(\xi). \tag{2.17}$$

Note that although we shall ultimately be interested in fields defined on space–time, for the present didactic purposes it suffices to consider variables $\xi^i(t)$ that are functions only of time. The Euler–Lagrange equation that is implied by (2.17) reads

$$f_{ij}(\xi)\dot{\xi}^j = \frac{\partial H(\xi)}{\partial \xi^i}, \tag{2.18}$$

where

$$f_{ij}(\xi) = \frac{\partial a_j(\xi)}{\partial \xi^i} - \frac{\partial a_i(\xi)}{\partial \xi^j}. \tag{2.19}$$

The first term in (2.17) determines the canonical 1-form: $a_i(\xi)\dot{\xi}^i\,dt = a_i(\xi)\,d\xi^i$, while f_{ij} gives the symplectic 2-form: $da_i(\xi)\,d\xi^i = \frac{1}{2}f_{ij}(\xi)\,d\xi^i\,d\xi^j$.

To set up a canonical formalism, we proceed directly. We *do not* make the frequently heard statement that "the canonical momenta $\partial L/\partial\dot{\xi}^i = a_i(\xi)$ are constrained to depend on the coordinates ξ," and we *do not* embark on Dirac's method for constrained systems.

In fact, if the matrix f_{ij} possesses the inverse f^{ij} there are no constraints. Then (2.18) implies

$$\dot{\xi}^i = f^{ij}(\xi)\frac{\partial H(\xi)}{\partial \xi^j}. \tag{2.20}$$

When one wants to express this equation of motion by bracketing with the Hamiltonian

$$\dot{\xi}^i = \{H(\xi), \xi^i\} = \{\xi^j, \xi^i\}\frac{\partial H(\xi)}{\partial \xi^j}, \tag{2.21}$$

one is directly led to postulating the fundamental bracket as

$$\{\xi^i, \xi^j\} = -f^{ij}(\xi). \tag{2.22}$$

The Poisson bracket between functions of ξ is then defined by

$$\{F_1(\xi), F_2(\xi)\} = -\frac{\partial F_1(\xi)}{\partial \xi^i} f^{ij} \frac{\partial F_2(\xi)}{\partial \xi^j}. \tag{2.23}$$

One verifies that (2.22) satisfies the Jacobi identity by virtue of (2.19).

When f_{ij} is singular and has no inverse, constraints do arise, and the development becomes more complicated (see [5]).

Our problem in connection with (2.14) and (2.16) is, in fact, the inverse of what we have here summarized. From (2.14) and (2.16), we know the form of f^{ij}, and that the Jacobi identity holds. We then wish to determine the inverse f_{ij}, and also a_i from (2.19). Since we know the Hamiltonian from (2.11), construction of the Lagrangian (2.17) should follow immediately.

However, an obstacle may arise: If there exists a quantity $C(\xi)$ whose Poisson bracket with all the ξ^i vanishes, then

$$0 = \{\xi^i, C(\xi)\} = -f^{ij} \frac{\partial}{\partial \xi^j} C(\xi). \tag{2.24}$$

That is, f^{ij} has the zero mode $\partial C(\xi)/\partial \xi^j$, and the inverse to f^{ij}, namely the symplectic 2-form f_{ij}, does not exist. In that case, something more has to be done, and we shall come back to this problem.

Totally commuting quantities like $C(\xi)$ are called "Casimir invariants." Since they Poisson-commute with *all* the dynamical variables, they commute with the Hamiltonian, and are constants of motion. But these constants do not reflect any symmetry of the specific Hamiltonian, nor do they generate any infinitesimal transformation on ξ^i, since the $\{C(\xi), \xi^i\}$ bracket vanishes.

As will be demonstrated below, the algebra (2.14), (2.16) admits Casimir invariants, which create an obstruction to the construction of a canonical formalism for fluid mechanics; this obstruction must be overcome to make progress. (In the Lagrangian formulation of fluid mechanics these Casimirs are related to a parametrization-invariance of that formalism [3].)

2.3 The Irrotational Case

We now return to the specific issue of determining the fluid dynamical Lagrangian. The problem of constructing a Lagrangian which leads to (2.14) and (2.16) can be solved by inspection for the irrotational case, with vanishing vorticity [see (2.6)]. For then the velocity commutator in (2.14) vanishes and (2.7) shows that the first equation in (2.14) can be satisfied by taking ρ and θ to be canonically conjugate

$$\{\theta(\boldsymbol{r}), \rho(\boldsymbol{r}')\} = \delta(\boldsymbol{r} - \boldsymbol{r}'). \tag{2.25}$$

Thus the Lagrangian reads

$$L_{\text{irrotational}} = \int dr\, (\theta\dot{\rho} - H_{\boldsymbol{v}=\boldsymbol{\nabla}\theta}), \tag{2.26}$$

where H is given by (2.11) with $\boldsymbol{v}$ taken as in (2.7). The form of this Lagrangian can be understood by the following argument, due to C. Eckart [6].

Consider the Lagrangian for N point-particles in free non-relativistic motion. With the mass m set to unity, the Galileo-invariant, free Lagrangian is just the kinetic energy

$$L_0 = \frac{1}{2}\sum_{n=1}^{N} v_n^2(t). \tag{2.27}$$

In a continuum description, the particle-counting index n becomes the continuous variable $\boldsymbol{r}$, and the particles are distributed with density ρ, so that $\sum_{n=1}^{N} v_n^2(t)$ becomes $\int dr\, \rho(t,\boldsymbol{r})v^2(t,\boldsymbol{r})$. But we also need to link the density with the current $\boldsymbol{j} = \rho\boldsymbol{v}$, so that the continuity equation holds. This can be enforced with the help of a Lagrange multiplier θ. We thus arrive at the free, continuum Lagrangian

$$\bar{L}_0^{\text{Galileo}} = \int dr \Big(\tfrac{1}{2}\rho v^2 + \theta\big(\dot{\rho} + \boldsymbol{\nabla}\cdot(\rho\boldsymbol{v})\big)\Big). \tag{2.28}$$

Since $\bar{L}_0^{\text{Galileo}}$ is first-order in time and the canonical 1-form $\int dr\, \theta\dot{\rho}$ does not contain $\boldsymbol{v}$, the latter may be varied, evaluated, and eliminated [5]. Doing this, we find

$$\rho \boldsymbol{v} - \rho \boldsymbol{\nabla} \theta = 0, \tag{2.29}$$

and we conclude that $\boldsymbol{\nabla}\theta$ is the velocity or, more precisely, that $\boldsymbol{\nabla}\theta$ is the $\boldsymbol{v}$ derivative of the kinetic energy, that is, the momentum $\boldsymbol{p}$, which in this nonrelativistic setting coincides with $\boldsymbol{v}$:

$$\boldsymbol{\nabla}\theta = \frac{\partial}{\partial \boldsymbol{v}} \tfrac{1}{2} v^2 \equiv \boldsymbol{p} = \boldsymbol{v}. \tag{2.30}$$

Substituting this in (2.28), we obtain

$$L_0^{\text{Galileo}} = \int dr \big(\theta \dot{\rho} - \tfrac{1}{2}\rho(\boldsymbol{\nabla}\theta)^2\big), \tag{2.31}$$

which reproduces (2.26) with the interaction $V(\rho)$ in (2.11) set to zero, and leads to the free version of the Bernoulli equation of motion (2.8)

$$\dot{\theta} + \frac{(\boldsymbol{\nabla}\theta)^2}{2} = 0. \tag{2.32}$$

Taking the gradient gives

$$\dot{\boldsymbol{v}} + \boldsymbol{v} \cdot \boldsymbol{\nabla} \boldsymbol{v} = 0. \tag{2.33}$$

Problem 2.3. The Lagrange density for the unit-mass Schrödinger equation can be taken as $\mathcal{L}_{\text{Schrödinger}} = i\hbar\psi^* \partial/\partial t \psi - \hbar^2/2m \boldsymbol{\nabla}\psi^* \cdot \boldsymbol{\nabla}\psi$. What form does this take after ψ is represented by $\rho^{1/2} e^{i\theta/\hbar}$? Compare with (2.26).

Remarkably, the same equation (2.33) emerges for a kinetic energy $T(\boldsymbol{v})$ that is an arbitrary function of $\boldsymbol{v}$. This will be useful for us when we study a relativistic generalization of the theory. If we replace (2.28) by

$$\bar{L}_0 = \int dr \Big(\rho T(\boldsymbol{v}) + \theta\big(\dot{\rho} + \boldsymbol{\nabla} \cdot (\rho \boldsymbol{v})\big)\Big), \tag{2.34}$$

and vary $\boldsymbol{v}$ to eliminate it, we get, in a generalization of (2.30),

$$\boldsymbol{\nabla}\theta = \frac{\partial T}{\partial \boldsymbol{v}} \equiv \boldsymbol{p}. \tag{2.35}$$

So, in the general case, it is the momentum—the $\boldsymbol{v}$ derivative of $T(\boldsymbol{v})$—that is irrotational. The Lagrange density becomes

$$L_0 = \int dr \left(\theta\dot{\rho} - \rho h(\boldsymbol{p})_{\boldsymbol{p}=\boldsymbol{\nabla}\theta}\right), \tag{2.36}$$

where $h(\boldsymbol{p})$ is the Legendre transform of $T(\boldsymbol{v})$:

$$h(\boldsymbol{p}) = \boldsymbol{v}\cdot\boldsymbol{p} - T(\boldsymbol{v}), \qquad \frac{\partial h}{\partial \boldsymbol{p}} = \boldsymbol{v}. \tag{2.37}$$

Again, varying θ in (2.36) gives the continuity equation

$$\begin{aligned} 0 = \frac{\delta L_0}{\delta\theta} &= \dot{\rho} - \int dr\, \rho \frac{\partial h(\boldsymbol{p})}{\partial \boldsymbol{p}} \cdot \frac{\delta \boldsymbol{p}}{\delta\theta} \\ &= \dot{\rho} - \int dr\, \rho\boldsymbol{v} \cdot \frac{\delta}{\delta\theta}\boldsymbol{\nabla}\theta \\ &= \dot{\rho} + \boldsymbol{\nabla}\cdot(\rho\boldsymbol{v}). \end{aligned} \tag{2.38}$$

Varying ρ gives

$$0 = \frac{\delta L_0}{\delta\rho} = -\dot{\theta} - h(\boldsymbol{p}). \tag{2.39}$$

Taking the gradient, this implies, with the help of (2.35),

$$\begin{aligned} \partial_i\dot{\theta} &= -\boldsymbol{v}\cdot\frac{\partial}{\partial r^i}\boldsymbol{p} \\ &= -v^j \frac{\partial}{\partial r^j} p^i \\ &= -v^j \frac{\partial p^i}{\partial v^k}\frac{\partial}{\partial r^j} v^k. \end{aligned} \tag{2.40}$$

On the other hand, (2.35) implies that

$$\partial_i\dot{\theta} = \frac{\partial p^i}{\partial v^k}\dot{v}^k. \tag{2.41}$$

The two are consistent, provided the free Euler equation holds, that is,

$$\dot{v}^k + v^j\,\partial_j v^k = 0 \tag{2.42}$$

(as long as $\partial p^i/\partial v^k = \partial^2 T/\partial v^i\,\partial v^k$ has an inverse).

Let us observe that free motion is here governed by a Lagrangian that is not quadratic and the free equations are not linear. Nevertheless, the equations of motion (2.38) and (2.42) can be solved in terms of initial data

$$\rho(t=0,\boldsymbol{r}) \equiv \rho_0(\boldsymbol{r}), \tag{2.43}$$

$$\boldsymbol{v}(t=0,\boldsymbol{r}) \equiv \boldsymbol{v}_0(\boldsymbol{r}). \tag{2.44}$$

Upon determining the retarded position $\boldsymbol{q}(t,\boldsymbol{r})$ from the equation

$$\boldsymbol{q} + t\boldsymbol{v}_0(\boldsymbol{q}) = \boldsymbol{r}, \tag{2.45}$$

one verifies that the solution to the free equations reads

$$\boldsymbol{v}(t,\boldsymbol{r}) = \boldsymbol{v}_0(\boldsymbol{q}), \tag{2.46}$$

$$\rho(t,\boldsymbol{r}) = \rho_0(\boldsymbol{q})\left|\det\frac{\partial q^i}{\partial r^j}\right|. \tag{2.47}$$

A final remark: Note that the free Bernoulli equation (2.8) coincides with the free Hamilton–Jacobi equation for the action.

2.4 Nonvanishing Vorticity and the Clebsch Parameterization

We now return to our original Galileo-invariant problem and enquire about the Lagrangian for velocity fields that are not irrotational, that is, whose vorticity is nonvanishing. Here we specify the spatial dimensionality to be 3, and observe that the algebra (2.14) possesses a zero mode, since the quantity

$$C(\boldsymbol{v}) \equiv \int dr\, \varepsilon^{ijk} v^i\, \partial_j v^k = \int dr\, \boldsymbol{v}\cdot\boldsymbol{\omega}, \tag{2.48}$$

(Poisson) commutes with both ρ and $\boldsymbol{v}$. So the symplectic 2-form does not exist: In the language developed above, f^{ij} has no inverse. (Notice that C, also called the "fluid helicity," coincides with the Abelian Chern–Simons term for $\boldsymbol{v}$ [7].) (In the irrotational case with vanishing $\boldsymbol{\omega}$, the obstacle obviously is absent.)

To make progress, one must neuteralize the obstruction. This is achieved in the following manner, as was shown by C.C. Lin [8].

We use the Clebsch parametrization for the vector field $\boldsymbol{v}$. Any three-dimensional vector, which involves three functions, can be presented as

$$\boldsymbol{v} = \boldsymbol{\nabla}\theta + \alpha\boldsymbol{\nabla}\beta, \tag{2.49}$$

with three suitably chosen scalar functions θ, α, and β. This is called the "Clebsch parametrization," and (α, β) are called "Gaussian potentials" [9]. In this parametrization, the vorticity reads

$$\boldsymbol{\omega} = \boldsymbol{\nabla}\alpha \times \boldsymbol{\nabla}\beta, \tag{2.50}$$

and the Lagrangian is taken as

$$L = -\int d\boldsymbol{r}\, \rho(\dot{\theta} + \alpha\dot{\beta}) - H_{\boldsymbol{v}=\boldsymbol{\nabla}\theta+\alpha\boldsymbol{\nabla}\beta}, \tag{2.51}$$

with $\boldsymbol{v}$ in H expressed as in (2.49). Thus (ρ, θ) remain canonically conjugate but another canonical pair appears: $(\rho\alpha, \beta)$. The phase space is four-dimensional, corresponding to the four observables ρ and $\boldsymbol{v}$, and a straightforward calculation shows that the Poisson brackets (2.14) are reproduced, with $\boldsymbol{v}$ constructed by (2.49).

But how has the obstacle been overcome? Let us observe that in the Clebsch parametrization C is given by

$$C = \int dr\, \varepsilon^{ijk} \partial_i\theta\, \partial_j\alpha\, \partial_k\beta, \tag{2.52}$$

which is just a surface integral

$$C = \int d\boldsymbol{S} \cdot (\theta\boldsymbol{\omega}). \tag{2.53}$$

In this form, C has no bulk contribution, and presents no obstacle to constructing a symplectic 2-form and a canonical 1-form in terms of θ, α, and β, which are defined in the bulk, that is, for all finite $\boldsymbol{r}$.

Lin gave an Eckart-type derivation of (2.51): Return to $\bar{L}_0^{\text{Galileo}}$ in (2.28) and add a further constraint, beyond the one enforcing current conservation [8]

$$\bar{L}_0^{\text{Galileo}} = \int dr \left(\tfrac{1}{2}\rho v^2 + \theta\big(\dot{\rho} + \nabla\cdot(\rho \boldsymbol{v})\big) - \rho\alpha\big(\dot{\beta} + \boldsymbol{v}\cdot\nabla\beta\big)\right). \quad (2.54)$$

Setting the variation (with respect to $\boldsymbol{v}$) to zero evaluates $\boldsymbol{v}$ as in (2.49); eliminating $\boldsymbol{v}$ from (2.54) gives rise to (2.51).

This procedure works in any number of dimensions, producing the same canonical 1-form in any dimension. This means that in two spatial dimensions, on the plane, where the $(\rho, \boldsymbol{v})$ space is three-dimensional, the four-dimensional phase space $(\rho, \theta; \rho\alpha, \beta)$ is larger. Moreover, the analog to C in two spatial dimensions, that is, the obstruction to constructing a symplectic 2-form, is not a single quantity: an infinite number of objects (Poisson) commute with ρ and $\boldsymbol{v}$. These are

$$C_n = \int dr\, \rho \left(\frac{\omega}{\rho}\right)^n, \qquad n = 0, \pm 1, \pm 2, \dots . \quad (2.55)$$

(Of course, the C_n vanish in the irrotational case where there is no obstruction.) One can understand why there is an infinite number of obstructions by observing that phase space must be even dimensional, but $(\rho, \boldsymbol{v})$ comprise three quantities on the plane. So a nonsingular symplectic form can be constructed either by increasing the number of canonical variables to four, or by decreasing to two. The Lin–Clebsch method increases the variables. On the other hand, the decreasing to two entails the suppressing of one continuous and local degree of freedom and, evidently, this is equivalent to neutralizing the infinite number of global obstructions, namely, the C_n. But we do not know how to effect such a suppression, so we remain with (2.51).

Note that $\bar{L}_0^{\text{Galileo}}$ in (2.54), apart from a total time derivative, can also be written in any number of dimensions as

$$\begin{aligned} \bar{L}_0^{\text{Galileo}} &= \int dr\, \big(\rho T(\boldsymbol{v}) - \rho(\dot{\theta} + \alpha\dot{\beta}) - \rho\boldsymbol{v}\cdot(\nabla\theta + \alpha\nabla\beta)\big) \\ &= \int dr\, \big(\rho T(\boldsymbol{v}) - j^\mu(\partial_\mu\theta + \alpha\partial_\mu\beta)\big), \end{aligned} \quad (2.56)$$

where we have introduced the current 4-vector

$$j^{\mu} = (c\rho, \rho \boldsymbol{v}), \tag{2.57}$$

employed the 4-vector gradient $\partial_{\mu} = (c^{-1}\,\partial/\partial t, \boldsymbol{\nabla})$, and denoted the kinetic energy by $T(\boldsymbol{v})$. These expressions will form our starting point for a relativistic generalization of the theory, as well as a non-Abelian generalization. (That is why we have introduced the velocity of light in the above definitions; of course, c disappears in the Galilean theory, as it has no role there.)

Finally, we observed that in one spatial dimension, where v can always be written as θ' and the vorticity vanishes, the irrotational canonical 1-form $\int dx\,\theta\dot{\rho}$ is generally applicable and can, equivalently, be written as $-\frac{1}{2}\int dx\,dy\,\rho(x)\varepsilon(x-y)\dot{v}(y)$, where ε is a ± 1-step function, determined by the sign of its argument. Evidently this leads to a spatially nonlocal but, otherwise , completely satisfactory canonical formulation of fluid motion on a line.

2.5 Some Further Remarks on the Clebsch Parameterization

Let us elaborate on the Clebsch parametrization for a vector field, which was presented for the velocity vector in (2.49). Here we shall use the notation of electromagnetism and discuss the Clebsch parametrization of a vector potential $\boldsymbol{A}$, which also leads to the magnetic field $\boldsymbol{B} = \boldsymbol{\nabla} \times \boldsymbol{A}$. (Of course the same observations apply when the vector in question is the velocity field $\boldsymbol{v}$, with $\boldsymbol{\nabla} \times \boldsymbol{v}$ giving the vorticity.)

The familiar parametrization of a three-component vector employs a scalar function θ (the "gauge" or "longitudinal" part) and a two-component transverse vector $\boldsymbol{A}_T$: $\boldsymbol{A} = \boldsymbol{\nabla}\theta + \boldsymbol{A}_T$, $\boldsymbol{\nabla} \cdot \boldsymbol{A}_T = 0$. This decomposition is unique and invertible (on a space with simple topology). In contrast, the Clebsch parametrization uses three scalar functions, θ, α, and β,

$$\boldsymbol{A} = \boldsymbol{\nabla}\theta + \alpha\boldsymbol{\nabla}\beta, \tag{2.58}$$

which are not uniquely determined by $\boldsymbol{A}$ (see below). The associated magnetic field reads

$$\boldsymbol{B} = \boldsymbol{\nabla}\alpha \times \boldsymbol{\nabla}\beta. \tag{2.59}$$

Repeating the above in form notation, the 1-form $A = A_i\, dr^i$ is presented as

$$A = d\theta + \alpha\, d\beta \tag{2.60}$$

and the 2-form is

$$dA = d\alpha\, d\beta\,. \tag{2.61}$$

Darboux's theorem ensures that the Clebsch parametrization is attainable locally in space [in the form (2.60)] [10]. Additionally, an explicit construction of α, β, and θ can be given by the following [11].

Solve the equations

$$\frac{dx}{B_x} = \frac{dy}{B_y} = \frac{dz}{B_z} \tag{2.62a}$$

which may also be presented as

$$\varepsilon^{ijk}\, dr^j\, B^k = 0. \tag{2.62b}$$

Solutions of these relations define two surfaces, called "magnetic surfaces," that are given by equations of the form

$$S_n(\boldsymbol{r}) = \text{constant} \qquad (n = 1, 2). \tag{2.63}$$

It follows from (2.62) that these also satisfy

$$\boldsymbol{B} \cdot \boldsymbol{\nabla} S_n = 0 \qquad (n = 1, 2), \tag{2.64}$$

that is, the normals to S_n are orthogonal to $\boldsymbol{B}$, or $\boldsymbol{B}$ is parallel to the tangent of S_n. The intersection of the two surfaces forms the so-called "magnetic lines," that is, loci that solve the dynamical system

$$\frac{d\boldsymbol{r}(\tau)}{d\tau} = \boldsymbol{B}\big(\boldsymbol{r}(\tau)\big), \tag{2.65}$$

where τ is an evolution parameter. Finally, the Gaussian potentials α and β are constructed as functions of $\boldsymbol{r}$ only through a dependence on the magnetic surfaces,

$$\begin{aligned} \alpha(\boldsymbol{r}) &= \alpha\big(S_n(\boldsymbol{r})\big), \\ \beta(\boldsymbol{r}) &= \beta\big(S_n(\boldsymbol{r})\big), \end{aligned} \tag{2.66}$$

so that

$$\boldsymbol{\nabla}\alpha \times \boldsymbol{\nabla}\beta = (\boldsymbol{\nabla}S_1 \times \boldsymbol{\nabla}S_2)\varepsilon^{mn}\frac{\partial\alpha}{\partial S_m}\frac{\partial\beta}{\partial S_n}. \tag{2.67}$$

Evidently, as a consequence of (2.64), $\boldsymbol{\nabla}\alpha \times \boldsymbol{\nabla}\beta$ in (2.67) is parallel to $\boldsymbol{B}$ and, because $\boldsymbol{B}$ is divergence-free, α and β can be adjusted so that the norm of $\boldsymbol{\nabla}\alpha \times \boldsymbol{\nabla}\beta$ coincides with $|\boldsymbol{B}|$. Once α and β have been fixed in this way, θ can easily be computed from $\boldsymbol{A} - \alpha\boldsymbol{\nabla}\beta$.

Neither the individual magnetic surfaces nor the Gauss potentials are unique. [By viewing A as a canonical 1-form, it is clear that expression (2.60) retains its form after a canonical transformation of α, β.] One may therefore require that the Gaussian potentials α and β simply coincide with the two magnetic surfaces: $\alpha = S_1$, $\beta = S_2$. Nevertheless, for a given $\boldsymbol{A}$ and $\boldsymbol{B}$, it may not be possible to solve (2.62) explicitly.

The Chern–Simons integrand $\boldsymbol{A} \cdot \boldsymbol{B}$ becomes, in the Clebsch parametrization,

$$\boldsymbol{A} \cdot \boldsymbol{B} = \boldsymbol{\nabla}\theta \cdot (\boldsymbol{\nabla}\alpha \times \boldsymbol{\nabla}\beta) = \boldsymbol{\nabla} \cdot (\theta\boldsymbol{B}) = \boldsymbol{B} \cdot \boldsymbol{\nabla}\theta. \tag{2.68}$$

Thus having identified the Gauss potentials α and β with the two magnetic surfaces, we deduce from (2.64) and (2.68) three equations for the three functions $(\theta,\ \alpha,\ \beta)$ that comprise the Clebsch parametrization

$$\begin{aligned} &\boldsymbol{B} \cdot \boldsymbol{\nabla}\alpha = \boldsymbol{B} \cdot \boldsymbol{\nabla}\beta = 0, \\ &\boldsymbol{B} \cdot \boldsymbol{\nabla}\theta = \text{Chern–Simons density } \boldsymbol{A} \cdot \boldsymbol{B}. \end{aligned} \tag{2.69}$$

Equation (2.68) also shows that in the Clebsch parametrization the Chern–Simons density becomes a total derivative

$$\boldsymbol{A} \cdot \boldsymbol{B} = \boldsymbol{\nabla} \cdot (\theta\boldsymbol{B}). \tag{2.70}$$

This does *not* mean that the Clebsch parametrization is unavailable when the Chern–Simons integral over all space is nonzero. Rather, for a nonvanishing integral and well-behaved $\boldsymbol{B}$ field, one must conclude that the Clebsch function θ is singular either in the finite volume of the integration region or on the surface at infinity bounding the integration domain. Then the Chern–Simons volume integral over (Ω) becomes a surface integral on the surfaces $(\partial\Omega)$ bounding the singularities

$$\int_\Omega dr\, \boldsymbol{A}\cdot\boldsymbol{B} = \int_{\partial\Omega} d\boldsymbol{S}\cdot(\theta\boldsymbol{B}). \tag{2.71}$$

Equation (2.71) shows that the Chern–Simons integral measures the magnetic flux, modulated by θ and passing through the surfaces that surround the singularities of θ.

The following explicit example illustrates the above points.

Consider the vector potential whose spherical components are given by

$$\begin{aligned} A_r &= (\cos\Theta)a'(r), \\ A_\Theta &= -(\sin\Theta)\frac{1}{r}\sin a(r), \\ A_\Phi &= -(\sin\Theta)\frac{1}{r}\big(1-\cos a(r)\big). \end{aligned} \tag{2.72}$$

(r, and Θ, Φ denote the conventional radial coordinate and the polar, azimuthal angles.) The function $a(r)$ is taken to vanish at the origin, and to behave as $2\pi\nu$ at infinity (ν integer or half-integer). The corresponding magnetic field reads

$$\begin{aligned} B_r &= -2(\cos\Theta)\frac{1}{r^2}\big(1-\cos a(r)\big), \\ B_\Theta &= (\sin\Theta)\frac{1}{r}a'(r)\sin a(r), \\ B_\Phi &= (\sin\Theta)\frac{1}{r}a'(r)\big(1-\cos a(r)\big), \end{aligned} \tag{2.73}$$

and the Chern–Simons integral—also called the "magnetic helicity" in the electrodynamical context—is quantized (in multiples of $16\pi^2$) by the behavior of $a(r)$ at infinity

$$\int d\boldsymbol{r}\, \boldsymbol{A}\cdot\boldsymbol{B} = -8\pi \int_0^\infty dr\, \frac{d}{dr}\big(a(r) - \sin a(r)\big)$$
$$= -16\pi^2\nu. \tag{2.74}$$

In spite of the nonvanishing magnetic helicity, a Clebsch parametrization for (2.72) is readily constructed. In form notation, it reads

$$A = d(-2\Phi) + 2\left(1 - \left(\sin^2\frac{a}{2}\right)\sin^2\Theta\right) \times d\left(\Phi + \tan^{-1}\left[\left(\tan\frac{a}{2}\right)\cos\Theta\right]\right). \tag{2.75}$$

The magnetic surfaces can be taken from formula (2.75) to coincide with the Gauss potentials

$$S_1 = 2\left(1 - \left(\sin^2\frac{a}{2}\right)\sin^2\Theta\right) = \text{constant},$$
$$S_2 = \Phi + \tan^{-1}\left[\left(\tan\frac{a}{2}\right)\cos\Theta\right] = \text{constant}. \tag{2.76}$$

The magnetic lines are determined by the intersection of S_1 and S_2:

$$\cos\frac{a}{2} = \varepsilon\cos(\Phi - \varphi_0),$$
$$\sin\Theta = \sqrt{\frac{1-\varepsilon^2}{1-\varepsilon^2\cos^2(\Phi-\varphi_0)}}, \tag{2.77}$$

where ε and φ_0 are constants. The potential $\theta = -2\Phi$ is multivalued. Consequently, the "surface" integral determining the Chern–Simons term reads

$$\int d\boldsymbol{r}\, \boldsymbol{A}\cdot\boldsymbol{B} = \int d\boldsymbol{r}\, \boldsymbol{\nabla}\cdot(-2\Phi\boldsymbol{B})$$
$$= \int_0^\infty r\, dr \int_0^\pi d\Theta\, B_\Phi|_{\Phi=2\pi}. \tag{2.78}$$

That is, the magnetic helicity is the flux of the toroidal magnetic field through the positive-x (x, z) half-plane.

Problem 2.4. Consider a vector potential $\boldsymbol{A}$, whose Clebsch parametrization reads $A_i = \partial_i \Phi + \cos\Theta\, \partial_i h(r)$, where Θ and Φ are the azimuthal and polar angles of the vector $\boldsymbol{r}$, and h is a nonsingular function of the magnitude of $\boldsymbol{r}$. Show that the Chern–Simons density (magnetic helicity density) is given by $\varepsilon^{ijk} A_i\, \partial_j A_k = \varepsilon^{ijk}\, \partial_i \Phi \partial_j \cos\Theta\, \partial_k h(r)$. Consider the integral of the Chern–Simons density over all space. This integral may first be evaluated over a spherical ball Ω, and then the radius R of the ball is taken to infinity. When the integrand is a divergence of a vector, Gauss's theorem casts the volume integral onto a surface integral over the sphere $\partial\Omega$ bounding the ball:

$$I = \int_\Omega d^3 r\, \boldsymbol{\nabla} \cdot \boldsymbol{V} = \int_{\partial\Omega} d\boldsymbol{S} \cdot \boldsymbol{V} = \int_0^{2\pi} d\Phi \int_0^\pi d\Theta \sin\Theta r^2 V^r|_{r=R},$$

but singularities in $\boldsymbol{V}$ may modify the equality. The three derivatives in the above Chern–Simons density may be extracted in three different ways. Show that the result is always $4\pi[h(R) - h(0)]$, but various singularities must be carefully handled.

Problem 2.5. Show that $\int d^3 r\, \boldsymbol{B} \cdot \delta\boldsymbol{A}$ vanishes (apart from surface terms) where $\delta\boldsymbol{A}$ is a variation and $\boldsymbol{A}$, $\boldsymbol{B} = \boldsymbol{\nabla} \times \boldsymbol{A}$, as well as $\delta\boldsymbol{A}$, are presented in the Clebsch parametrization. When the variational principle is implemented, by varying the components of $\boldsymbol{A}$, one finds that $\frac{1}{2}\int d^3 r\, B^2$ is stationary provided $\boldsymbol{\nabla} \times \boldsymbol{B} = 0$. Show that implementing the variation, by varying the scalar functions in the Clebsch parametrization for $\boldsymbol{A}$, gives the weaker condition $\boldsymbol{\nabla} \times \boldsymbol{B} = \mu \boldsymbol{B}$, where μ can depend on $\boldsymbol{r}$. How is this $\boldsymbol{r}$-dependence constrained? How is the constraint satisfied?

There is another approach to the construction of (Abelian) vector potentials for which the (Abelian) Chern–Simons density is a total derivative and, as a consequence, a Clebsch parametrization for these potentials is readily found. The method relies on projecting an Abelian potential from a non-Abelian one, and it can be generalized to a construction of non-Abelian vectors for which the non-Abelian Chern–Simons density is again a total derivative. This will be useful for us when we come to discuss non-Abelian fluid mechanics. Therefore, this method will now be explained—in its Abelian realization [12].

We consider an SU(2) group element g and a pure gauge SU(2) gauge field, whose matrix-valued 1-form is

$$g^{-1}\,dg = V^a \frac{\sigma^a}{2i}, \tag{2.79}$$

where σ^a are Pauli matrices. It is known that

$$\mathrm{tr}(g^{-1}\,dg)^3 = -\tfrac{1}{4}\varepsilon^{abc}V^aV^bV^c = -\tfrac{3}{2}V^1V^2V^3, \tag{2.80}$$

is a total derivative; indeed its spatial integral measures the winding number of the gauge function g [13]. Since V^a is a pure gauge, we have

$$dV^a = -\tfrac{1}{2}\varepsilon^{abc}V^bV^c, \tag{2.81}$$

so that if we define an Abelian gauge potential A by projecting one SU(2) component of (2.79) (say the third) $A = V^3$, the Abelian Chern–Simons density for A is a total derivative, as is seen from the chain of equations that relies on (2.80) and (2.81):

$$A\,dA = V^3\,dV^3 = -V^1V^2V^3 = \tfrac{2}{3}\,\mathrm{tr}(g^{-1}\,dg)^3. \tag{2.82}$$

Of course, $A = V^3$ is not an Abelian pure gauge.

Note that g depends on three arbitrary functions, the three SU(2) local gauge functions. Hence V^3 enjoys sufficient generality to represent the three-dimensional vector A. Moreover, since A's Abelian Chern–Simons density is given by $\mathrm{tr}(g^{-1}\,dg)^3$, which is a total derivative, a Clebsch parametrization for A is easily constructed. We also observe that when the SU(2) group element g has a nonvanishing winding number, the resultant Abelian vector possesses a nonvanishing Chern–Simons integral, that is, nonzero magnetic helicity.

Final,y we remark that the example of a Clebsch-parametrized gauge potential A, presented above in (2.72), is obtained by a projection onto the third isospin direction of a pure gauge SU(2) potential, constructed from group element $g = \exp\big((\sigma^a/2i)\hat{r}^a a(r)\big)$ [12].

Further intricacies arise when the Clebsch parametrization is used in variational calculations involving vector fields; see [14].

3

Specific Models

We now return to our irrotational models, both relativistic and nonrelativistic, for which we shall specify an explicit force law and discuss further properties.

3.1 Galileo-Invariant Nonrelativistic Model

Recall that the nonrelativistic Lagrangian for irrotational motion reads

$$L^{\text{Galileo}} = \int dr \left(\theta \dot{\rho} - \rho \frac{(\boldsymbol{\nabla}\theta)^2}{2} - V(\rho) \right), \tag{3.1}$$

where $\boldsymbol{\nabla}\theta = \boldsymbol{v}$. The Hamiltonian density $\mathcal{H}$ is composed of the last two terms beyond the canonical 1-form $\int dr\, \theta \dot{\rho}$:

$$H = \int dr \left(\rho \frac{(\boldsymbol{\nabla}\theta)^2}{2} + V(\rho) \right) = \int dr\, \mathcal{H}. \tag{3.2}$$

Various expressions for V appear in the literature. $V(\rho) \propto \rho^n$ is a popular choice, appropriate for the adiabatic equation of state. We shall be specifically interested in the "Chaplygin gas."

$$V(\rho) = \frac{\lambda}{\rho}, \qquad \lambda > 0. \tag{3.3}$$

According to what we said before, the Chaplygin gas has enthalpy $V' = -\lambda/\rho^2$, negative pressure $P = -2\lambda/\rho$, and speed of sound $s = \sqrt{2\lambda}/\rho$ (hence $\lambda > 0$).

Chaplygin introduced his equation of state as a mathematical approximation to the physically relevant adiabatic expressions with $n > 0$. (Constants are arranged so that the Chaplygin formula is tangent at one point to the adiabatic profile [15].) Also, it was realized that certain deformable solids can be described by the Chaplygin equation of state [16]. At present, negative pressure is recognized as a possible physical effect: exchange forces in atoms give rise to negative pressure; stripe states in the quantum Hall effect may be a consequence of negative pressure; and the recently discovered cosmological constant may be exerting negative pressure on the cosmos, thereby accelerating expansion.

For any form of V, the model possesses the Galileo symmetry appropriate to nonrelativistic dynamics. The Galileo transformations comprise time and space translations, as well as space rotations. The corresponding constants of motion are the energy E:

$$E = \int dr\, \mathcal{H} = \int dr \left(\rho \frac{(\boldsymbol{\nabla}\theta)^2}{2} + V(\rho) \right) \quad \text{(time translation)}, \tag{3.4}$$

the momentum $\boldsymbol{P}$ (whose density $\boldsymbol{\mathcal{P}}$ equals the spatial current):

$$\boldsymbol{P} = \int dr\, \boldsymbol{\mathcal{P}} = \int dr\, \boldsymbol{j} = \int dr\, \rho \boldsymbol{v} \quad \text{(space translation)}, \tag{3.5}$$

and the angular momentum M^{ij}:

$$M^{ij} = \int dr (r^i \mathcal{P}^j - r^j \mathcal{P}^i) \qquad \text{(rotation)}. \tag{3.6}$$

The action of these transformations on the fields is straightforward: the time and space arguments are shifted or the space argument is rotated. Slightly less trivial is the action of Galileo boosts, which boost the spatial coordinate by a velocity $\boldsymbol{u}$:

$$\boldsymbol{r} \to \boldsymbol{R} = \boldsymbol{r} - t\boldsymbol{u}. \tag{3.7}$$

The density field transforms trivially: its spatial argument is boosted,

$$\rho(t, \boldsymbol{r}) \to \rho_{\boldsymbol{u}}(t, \boldsymbol{r}) = \rho(t, \boldsymbol{R}), \tag{3.8}$$

but the velocity potential acquires an inhomogeneous term:

$$\theta(t, \boldsymbol{r}) \to \theta_{\boldsymbol{u}}(t, \boldsymbol{r}) = \theta(t, \boldsymbol{R}) + \boldsymbol{u} \cdot \boldsymbol{r} - \frac{u^2}{2} t. \tag{3.9}$$

Those of you familiar with field-theoretic realizations of the Galileo group will recognize the inhomogeneous term as the well-known Galileo 1-cocyle. It ensures that the velocity, the gradient of θ, transforms appropriately,

$$\boldsymbol{v}(t,\boldsymbol{r}) \to \boldsymbol{v}_{\boldsymbol{u}}(t,\boldsymbol{r}) = \boldsymbol{v}(t,\boldsymbol{r}-t\boldsymbol{u}) + \boldsymbol{u}. \tag{3.10}$$

The associated conserved quantity is the "boost generator"

$$\boldsymbol{B} = t\boldsymbol{P} - \int dr\, \rho\boldsymbol{r} \qquad \text{(Galileo boost)}. \tag{3.11}$$

Finally, also conserved is the total number

$$N = \int dr\, \rho \qquad \text{(particle number)}. \tag{3.12}$$

The corresponding transformation shifts θ by a constant.

These transformations and constants of motion fit into the general theory: the action is invariant against the transformations; Noether's theorem can be used to derive the constants of motion; and their time independence is verified with the help of the equations of motion—indeed, the continuity equations (2.1), (2.4), and (2.5), as well as the symmetry of T^{ij}, guarantee this. Also, using the basic Poisson brackets (2.25) for the (ρ,θ) variables, one can check that each infinitesimal transformation is generated by Poisson bracketing with the appropriate constant; Poisson bracketing the constants with each other reproduces the Galileo–Lie algebra with a central extension given by N, which corresponds to the familiar Galileo 2-cocycle. There are a total of $\frac{1}{2}(d+1)(d+2)$ Galileo generators in d space plus one time dimension. Together with the central term, we have a total of $\frac{1}{2}(d+1)(d+2)+1$ generators.

Another useful consequence of the symmetry transformations is that they map solutions of the equations of motion into new solutions. Of course, "new" solutions produced by Galileo transformations are trivially related to the old ones: they are simply shifted, boosted, or rotated.

[The free theory, as well as the adiabatic theory with $n = 1+2/d$, are also invariant against the SO(2, 1) group of time translation, dilation, and conformal transformation [17], which together with the Galileo group form the "Schrödinger group" of nonrelativistic motion, whenever the energy-momentum "tensor" satisfies $2T^{00} = T^i_i$ [18].]

But we shall now turn to the specific Chaplygin gas model, with $V(\rho) = \lambda/\rho$, which possesses additional and unexpected symmetries.

The Chaplygin gas action and consequent Bernoulli equation for the Chaplygin gas in $(d, 1)$ space–time read

$$I_\lambda^{\text{Chaplygin}} = \int dt \int dr \left(\theta\dot{\rho} - \tfrac{1}{2}\rho(\boldsymbol{\nabla}\theta)^2 - \frac{\lambda}{\rho} \right), \tag{3.13}$$

$$\dot{\theta} + \frac{(\boldsymbol{\nabla}\theta)^2}{2} = \frac{\lambda}{\rho^2}. \tag{3.14}$$

This model possesses further space–time symmetries beyond those of the Galileo group [19]. First of all, there is a one-parameter (ω) time rescaling transformation

$$t \to T = e^\omega t, \tag{3.15}$$

under which the fields transform as

$$\theta(t, \boldsymbol{r}) \to \theta_\omega(t, \boldsymbol{r}) = e^\omega \theta(T, \boldsymbol{r}), \tag{3.16}$$

$$\rho(t, \boldsymbol{r}) \to \rho_\omega(t, \boldsymbol{r}) = e^{-\omega} \rho(T, \boldsymbol{r}). \tag{3.17}$$

Second, in d spatial dimensions, there is a vectorial, d-parameter $(\boldsymbol{\omega})$ space–time mixing transformation

$$t \to T(t, \boldsymbol{r}) = t + \boldsymbol{\omega} \cdot \boldsymbol{r} + \tfrac{1}{2}\omega^2 \theta(T, \boldsymbol{R}), \tag{3.18}$$

$$\boldsymbol{r} \to \boldsymbol{R}(t, \boldsymbol{r}) = \boldsymbol{r} + \boldsymbol{\omega}\theta(T, \boldsymbol{R}). \tag{3.19}$$

Note that the transformation law for the coordinates involves the θ field itself. Under this transformation, the fields transform according to

$$\theta(t, \boldsymbol{r}) \to \theta_{\boldsymbol{\omega}}(t, \boldsymbol{r}) = \theta(T, \boldsymbol{R}), \tag{3.20}$$

$$\rho(t, \boldsymbol{r}) \to \rho_{\boldsymbol{\omega}}(t, \boldsymbol{r}) = \rho(T, \boldsymbol{R}) \frac{1}{|J|}, \tag{3.21}$$

with J the Jacobian of the transformation linking $(T, \boldsymbol{R}) \rightarrow (t, \boldsymbol{r})$:

$$J = \det \begin{pmatrix} \partial T/\partial t & \partial T/\partial r^j \\ \partial R^i/\partial t & \partial R^i/\partial r^j \end{pmatrix}$$
$$= \left(1 - \boldsymbol{\omega} \cdot \boldsymbol{\nabla}\theta(T, \boldsymbol{R}) - \frac{\omega^2}{2}\dot{\theta}(T, \boldsymbol{R})\right)^{-1}. \tag{3.22}$$

(The time and space derivatives in the last element are with respect to t and $\boldsymbol{r}$.) One can again tell the complete story for these transformations: the action is invariant; and Noether's theorem gives the conserved quantities which, for the time rescaling, is

$$D = tH - \int dr\, \rho\theta \qquad \text{(time rescaling)}, \tag{3.23}$$

while for the space–time mixing one finds

$$\boldsymbol{G} = \int dr\, (\boldsymbol{r}\mathcal{H} - \theta\boldsymbol{\mathcal{P}}) \qquad \text{(space–time mixing)}. \tag{3.24}$$

The time independence of D and $\boldsymbol{G}$ can be verified with the help of the equations of motion (continuity and Bernoulli). Poisson bracketing the fields θ and ρ with D and $\boldsymbol{G}$ generates the appropriate infinitesimal transformation on the fields.

So now the total number of generators is the sum of the previous $\frac{1}{2}(d+1)(d+2)+1$ with $1+d$ additional ones

$$\tfrac{1}{2}(d+1)(d+2) + 1 + 1 + d = \tfrac{1}{2}(d+2)(d+3). \tag{3.25}$$

When one computes the Poisson brackets of all these with each other, one finds the Poincaré–Lie algebra in one higher spatial dimension, that is, in $(d+1, 1)$-dimensional space–time, where the Poincaré group possesses $\frac{1}{2}(d+2)(d+3)$ generators [20]. Moreover, one verifies that $(t, \theta, \boldsymbol{r})$ transform linearly as a $(d+2)$ Lorentz vector in light-cone components, with t being the "+" component and θ the "−" component [21].

Presently, we shall use these additional symmetries to generate new solutions from old ones but, in contrast with what we saw earlier, the new solutions will be nontrivially linked to the former ones. Note that the additional symmetry holds even in the free theory.

Before proceeding, let us observe that ρ may be eliminated by using the Bernoulli equation to express it in terms of θ. In this way, one is led to the following ρ-independent action for θ in the Chaplygin gas problem:

$$I_\lambda^{\text{Chaplygin}} = -2\sqrt{\lambda}\int dt \int dr \sqrt{\dot{\theta} + \frac{(\boldsymbol{\nabla}\theta)^2}{2}}. \tag{3.26}$$

Although this operation is possible only in the interacting case, the interaction strength is seen to disappear from the equations of motion:

$$\frac{\partial}{\partial t}\frac{1}{\sqrt{\dot{\theta} + (\boldsymbol{\nabla}\theta)^2/2}} + \boldsymbol{\nabla}\cdot\frac{\boldsymbol{\nabla}\theta}{\sqrt{\dot{\theta} + (\boldsymbol{\nabla}\theta)^2/2}} = 0, \tag{3.27}$$

λ merely serves as an overall factor in the action.

The action (3.26) looks unfamiliar; yet it is Galileo invariant. [The combination $\dot{\theta} + \frac{1}{2}(\boldsymbol{\nabla}\theta)^2$ responds to Galileo transformations without a 1-cocycle; see (3.9).] Also $I_\lambda^{\text{Chaplygin}}$ possesses the additional symmetries described above, with θ transforming according to the previously recorded equations.

Let us discuss some solutions. For example, the free theory is solved by

$$\theta(t, \boldsymbol{r}) = \frac{r^2}{2t}, \tag{3.28}$$

which corresponds to the velocity

$$\boldsymbol{v}(t, \boldsymbol{r}) = \frac{\boldsymbol{r}}{t}. \tag{3.29}$$

Galileo transforms generalize this in an obvious manner into a set of solutions. (The charge density ρ is determined by its initial condition. In the free theory, ρ is an independent quantity, and we shall not discuss it here.) Performing on the above formula for θ, the new transformations of time-rescaling and space–time mixing, we find that the solution is invariant.

We can find a solution similar to (3.28) in the interacting case, for $d > 1$, which we henceforth assume (the $d = 1$ case will be discussed separately later). One verifies that a solution is

$$\begin{aligned} \theta(t,\boldsymbol{r}) &= -\frac{r^2}{2(d-1)t}, & \rho(t,\boldsymbol{r}) &= \sqrt{\frac{2\lambda}{d}}(d-1)\frac{|t|}{r}, \\ \boldsymbol{v}(t,\boldsymbol{r}) &= -\frac{\boldsymbol{r}}{(d-1)t}, & \boldsymbol{j}(t,\boldsymbol{r}) &= -\varepsilon(t)\sqrt{\frac{2\lambda}{d}}\hat{r}. \end{aligned} \tag{3.30}$$

Note that the speed of sound

$$s = \frac{\sqrt{2\lambda}}{\rho} = \frac{\sqrt{d}\,r}{(d-1)t} = \sqrt{d}v, \tag{3.31}$$

exceeds v. Again this solution can be translated, rotated, and boosted. Moreover, the solution is time-rescaling invariant. However, the space–time mixing transformation produces a wholly different kind of solution. This is best shown graphically, where the $d = 2$ case is exhibited (see Figure 3.1) [22].

Another interesting solution, which is essentially one-dimensional (lineal), even though it exists in arbitrary spatial dimension, is given by

$$\theta(t,\boldsymbol{r}) = \Theta(\hat{n}\cdot\boldsymbol{r}) + \boldsymbol{u}\cdot\boldsymbol{r} - \tfrac{1}{2}t\big(u^2 - (\hat{n}\cdot\boldsymbol{u})^2\big). \tag{3.32}$$

Here $\hat{n}$ is a spatial unit vector, $\boldsymbol{u}$ is an arbitrary vector with dimension of velocity, while Θ is an arbitrary function with static argument, which can be boosted by the Galileo transform (3.9). The corresponding charge density is time-independent

$$\rho(t,\boldsymbol{r}) = \frac{\sqrt{2\lambda}}{\hat{n}\cdot\boldsymbol{u} + \Theta'(\hat{n}\cdot\boldsymbol{r})}. \tag{3.33}$$

The current is static and divergenceless

$$\boldsymbol{j}(t,\boldsymbol{r}) = \sqrt{2\lambda}\left(\hat{n} + \frac{\boldsymbol{u} - \hat{n}(\hat{n}\cdot\boldsymbol{u})}{\hat{n}\cdot\boldsymbol{r} + \Theta'(\hat{n}\cdot\boldsymbol{r})}\right). \tag{3.34}$$

The sound speed $s = \sqrt{2\lambda}/\rho = \Theta'(\hat{n}\cdot\boldsymbol{r}) + \hat{n}\cdot\boldsymbol{u}$ is just the $\hat{n}$ component of the velocity $\boldsymbol{v} = \boldsymbol{\nabla}\theta = \hat{n}\Theta'(\hat{n}\cdot\boldsymbol{r}) + \boldsymbol{u}$.

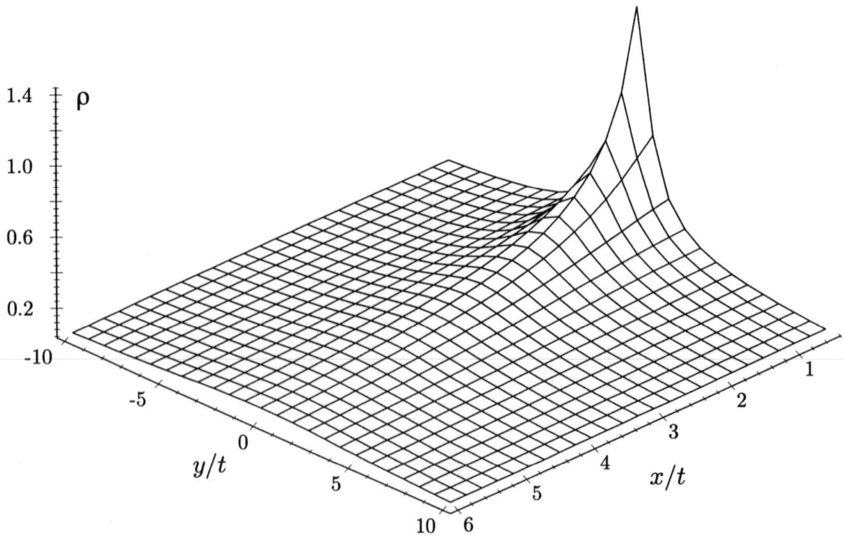

(a) The original density $\rho(t, \boldsymbol{r}) \propto |t|/r$ (in two dimensions, $\boldsymbol{r} = (x, y)$).

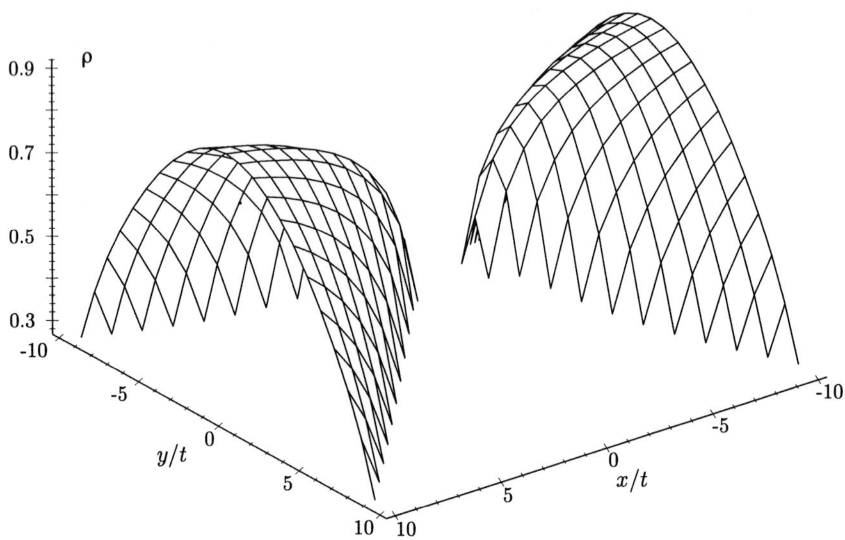

(b) The transformed density $\rho(t, \boldsymbol{r})$.

Finally, we record a planar static solution to (3.27)

$$\theta(\boldsymbol{r}) = \Theta(\hat{n}_1 \cdot \boldsymbol{r}/\hat{n}_2 \cdot \boldsymbol{r}), \tag{3.35}$$

where $\hat{n}_1$ and $\hat{n}_2$ are two orthogonal unit vectors [23].

Problem 3.1. Show that the solution for θ given in (3.28) is invariant under the time-rescaling transformation (3.15), (3.16), and under the space–time mixing transformation (3.18)–(3.20).

3.2 Lorentz-Invariant Relativistic Model

We now turn to a Lorentz-invariant generalization of our Galileo-invariant Chaplygin model in $(d, 1)$-dimensional space–time. We already know from (2.34)–(2.37) how to construct the free Lagrangian by using Eckart's method with a relativistic kinetic energy

$$T(\boldsymbol{v}) = -c^2\sqrt{1 - v^2/c^2}. \tag{3.36}$$

Recall that mass has been scaled to unity, and that we retain the velocity of light c to keep track of the nonrelativistic $c \to \infty$ limit. Evidently, the momentum is

$$\boldsymbol{p} = \frac{\partial T(\boldsymbol{v})}{\partial \boldsymbol{v}} = \frac{\boldsymbol{v}}{\sqrt{1 - v^2/c^2}}. \tag{3.37}$$

Thus the free relativistic Lagrangian, with current conservation enforced by the Lagrange multiplier θ, reads [compare (2.34)]

$$\bar{L}_0^{\text{Lorentz}} = \int dr \Big(-c^2 \rho \sqrt{1 - v^2/c^2} + \theta\big(\dot{\rho} + \boldsymbol{\nabla} \cdot (\rho \boldsymbol{v})\big) \Big). \tag{3.38}$$

This may be presented in a Lorentz-covariant form in terms of a current four-vector $j^\mu = (c\rho, \rho\boldsymbol{v})$. $\bar{L}_0^{\text{Lorentz}}$ of (3.38) is thus equivalent to [compare (2.56), (2.57)]

$$\bar{L}_0^{\text{Lorentz}} = \int dr \big(-j^\mu \partial_\mu \theta - c\sqrt{j^\mu j_\mu} \big). \tag{3.39}$$

Eliminating $\boldsymbol{v}$ in (3.38), we find, as before, that

$$\boldsymbol{p} = \frac{\partial T}{\partial \boldsymbol{v}} = \frac{\boldsymbol{v}}{\sqrt{1 - v^2/c^2}} = \boldsymbol{\nabla}\theta, \quad \boldsymbol{v} = \frac{\boldsymbol{\nabla}\theta}{\sqrt{1 + (\boldsymbol{\nabla}\theta)^2/c^2}}, \tag{3.40}$$

and the free Lorentz-invariant Lagrangian reads [compare (2.36), (2.37)]

$$L_0^{\text{Lorentz}} = \int dr \left(\theta\dot{\rho} - \rho c^2\sqrt{1 + (\boldsymbol{\nabla}\theta)^2/c^2}\right). \tag{3.41}$$

To find L_0^{Galileo} of (2.31) as the nonrelativistic limit of L_0^{Lorentz} in (3.41), a nonrelativistic θ variable must be extracted from its relativistic counterpart. Calling the former θ_{NR} and the latter, which occurs in (3.41), θ_{R}, we define

$$\theta_{\text{R}} \equiv -c^2 t + \theta_{\text{NR}}. \tag{3.42}$$

It then follows that, apart from a total time derivative,

$$L_0^{\text{Lorentz}} \xrightarrow{c\to\infty} L_0^{\text{Galileo}}. \tag{3.43}$$

Next, one wants to include interactions. While there are many ways to allow for Lorentz-invariant interactions, we seek an expression that reduces to the Chaplygin gas in the nonrelativistic limit. Thus, we choose [24]

$$L_a^{\text{Born–Infeld}} = \int dr \left(\theta\dot{\rho} - \sqrt{\rho^2 c^2 + a^2}\sqrt{c^2 + (\boldsymbol{\nabla}\theta)^2}\right), \tag{3.44}$$

where a is the interaction strength. [The reason for the nomenclature will emerge presently.] We see that, as $c \to \infty$,

$$L_a^{\text{Born–Infeld}} \xrightarrow{c\to\infty} L_{\lambda=a^2/2}^{\text{Chaplygin}}. \tag{3.45}$$

[Again θ_{NR} is extracted from θ_{R} as in (3.42) and a total time derivative is ignored.] Although it is perhaps not obvious, (3.44) defines a Poincaré-invariant theory, and this will be explicitly demonstrated below. Therefore, $L_a^{\text{Born–Infeld}}$ possesses Lorentz and Poincaré symmetries in $(d, 1)$ space–time, with a total of $\frac{1}{2}(d+1)(d+2)+1$ generators, where the last "+1" refers to the total number $N = \int dr\, \rho$.

When $a = 0$, the model is free and elementary. It was demonstrated previously [(2.34)–(2.42)] that the free equations of motion are precisely the same as in the nonrelativistic free model, so the complete solution (2.43)–(2.47) works here as well. For $a \neq 0$, in the presence of interactions, one can eliminate ρ as before, and one is left with a Lagrangian just for the θ field. It reads

$$L_a^{\text{Born–Infeld}} = -a \int dr\, \sqrt{c^2 - (\partial_\mu \theta)^2}. \tag{3.46}$$

This is a Born–Infeld-type theory for a scalar field θ; its Poincaré invariance is manifest and, again, the elimination of ρ is only possible with nonvanishing a which, however., disappears from the dynamics, serving merely to normalize the Lagrangian.

The equations of motion that follow from (3.44) read

$$\dot{\rho} + \boldsymbol{\nabla} \cdot \left(\boldsymbol{\nabla}\theta \sqrt{\frac{\rho^2 c^2 + a^2}{c^2 + (\boldsymbol{\nabla}\theta)^2}} \right) = 0, \tag{3.47}$$

$$\dot{\theta} + \rho c^2 \sqrt{\frac{c^2 + (\boldsymbol{\nabla}\theta)^2}{\rho^2 c^2 + a^2}} = 0. \tag{3.48}$$

The density ρ can be evaluated in terms of θ from (3.48); then (3.47) reads

$$\partial^\alpha \left(\frac{1}{\sqrt{c^2 - (\partial_\mu \theta)^2}}\, \partial_\alpha \theta \right) = 0, \tag{3.49}$$

which also follows from (3.46). After θ_{NR} is extracted from θ_{R}, as in (3.42), we see that in the nonrelativistic limit $L_a^{\text{Born–Infeld}}$ (3.44) or (3.46) becomes $L_\lambda^{\text{Chaplygin}}$ of (3.13) or (3.26),

$$L_a^{\text{Born–Infeld}} \xrightarrow{c \to \infty} L_{\lambda = a^2/2}^{\text{Chaplygin}}, \tag{3.50}$$

and the equations of motion (3.47)–(3.49) reduce to (2.1), (3.14), and (3.27).

In view of all the similarities to the nonrelativistic Chaplygin gas, it comes as no surprise that the relativistic Born–Infeld theory possesses additional symmetries. These additional symmetry transformations, which leave (3.44) or (3.46) invariant, involve a one-parameter (ω) reparametrization of time, and a d-parameter ($\boldsymbol{\omega}$) vectorial reparametrization of space. Both transformations are field dependent.

The time transformation is given by an implicit formula also involving thc ficld θ [25]

$$t \to T(t, \boldsymbol{r}) = \frac{t}{\cosh c^2\omega} + \frac{\theta(T, \boldsymbol{r})}{c^2} \tanh c^2\omega, \tag{3.51}$$

while the field transforms according to

$$\theta(t, \boldsymbol{r}) \to \theta_\omega(t, \boldsymbol{r}) = \frac{\theta(T, \boldsymbol{r})}{\cosh c^2\omega} - c^2 t \tanh c^2\omega. \tag{3.52}$$

[We record here only the transformation on θ; how ρ transforms can be determined from the (relativistic) Bernoulli equation, obtained by varying ρ in (3.44), which expresses ρ in terms of θ. Moreover, (3.52) is sufficient for discussing the invariance of (3.46).] The infinitesimal generator, which is time independent by virtue of the equations of motion, is [26]

$$\begin{aligned} D &= \int d\boldsymbol{r} \left(c^4 t\rho + \theta\sqrt{\rho^2c^2 + a^2}\sqrt{c^2 + (\boldsymbol{\nabla}\theta)^2}\right) \\ &= \int d\boldsymbol{r} (c^4 t\rho + \theta\mathcal{H}) \qquad \text{(time reparametrization)}. \end{aligned} \tag{3.53}$$

A second class of transformations involving a reparametrization of the spatial variables is implicitly defined by [25]

$$\begin{aligned} \boldsymbol{r} \to \boldsymbol{R}(t, \boldsymbol{r}) = \boldsymbol{r} - \widehat{\omega}\theta(t, \boldsymbol{R})\frac{\tan c\omega}{c} \\ + \hat{\omega}(\hat{\omega}\cdot\boldsymbol{r})\left(\frac{1 - \cos c\omega}{\cos c\omega}\right), \end{aligned} \tag{3.54}$$

$$\theta(t, \boldsymbol{r}) \to \theta_{\boldsymbol{\omega}}(t, \boldsymbol{r}) = \frac{\theta(t, \boldsymbol{R}) - c(\hat{\omega}\cdot\boldsymbol{r})\sin c\omega}{\cos c\omega}, \tag{3.55}$$

where $\widehat{\omega}$ is the unit vector $\boldsymbol{\omega}/\omega$ and $\omega = \sqrt{\boldsymbol{\omega}\cdot\boldsymbol{\omega}}$. The time-independent generator of the infinitesimal transformation reads [26]

$$\begin{aligned} \boldsymbol{G} &= \int d\boldsymbol{r} \left(c^2\boldsymbol{r}\rho + \theta\rho\boldsymbol{\nabla}\theta\right) \\ &= \int d\boldsymbol{r} \left(c^2\boldsymbol{r}\rho + \theta\boldsymbol{\mathcal{P}}\right) \qquad \text{(space reparametrization)}. \end{aligned} \tag{3.56}$$

Of course, the Born–Infeld action (3.44) or (3.46) is invariant against these transformations, whose infinitesimal form is generated by the constants.

With the addition of D and $\boldsymbol{G}$ to the previous generators, the Poincaré algebra in $(d+1,1)$ dimension is reconstructed, and $(t,\boldsymbol{r},\theta)$ transforms linearly as a $(d+2)$-dimensional Lorentz vector (in Cartesian components) [21]. Note that this symmetry also holds in the free, $a=0$, theory.

It is easy to exhibit solutions of the relativistic equation (3.49), which reduce to solutions of the nonrelativistic, Chaplygin gas equation (3.27) [after $-c^2 t$ has been removed, as in (3.42)]. For example,

$$\theta(t,\boldsymbol{r}) = -c\sqrt{c^2t^2 + \frac{r^2}{d-1}}, \tag{3.57}$$

solves (3.49) and reduces to (3.30). The relativistic analog of the lineal solution (3.32) is

$$\theta(t,\boldsymbol{r}) = \Theta(\hat{n}\cdot\boldsymbol{r}) + \boldsymbol{n}\cdot\boldsymbol{r} - ct\sqrt{c^2+u^2-(\hat{n}\cdot\boldsymbol{u})^2}. \tag{3.58}$$

[Note that the above profiles continue to solve (3.49), even when the sign of the square root is reversed, but then they no longer possess a nonrelativistic limit.]

Additionally there exists an essentially relativistic solution, describing massless propagation in one direction: according to (3.49), θ can satisfy the wave equation $\Box\theta=0$, provided $(\partial_\mu\theta)^2 =$ constant as, for example, with plane waves

$$\theta(t,\boldsymbol{r}) = f(\hat{n}\cdot\boldsymbol{r} \pm ct), \tag{3.59}$$

where $(\partial_\mu\theta)^2$ vanishes. Then ρ reads, from (3.48),

$$\rho = \mp\frac{a}{c^2}f'. \tag{3.60}$$

Other solutions are given in [27].

3.3 Some Remarks on Relativistic Fluid Mechanics

The Born–Infeld model reduces in the nonrelativistic limit to the Chaplygin gas. Equations governing the latter belong to fluid mechanics, but the Born–Infeld equations do not readily expose their fluid mechanical structure. Nevertheless, they do in fact describe a relativistic fluid. In order to demonstrate this, we give a précis of relativistic fluid mechanics.

Usually the dynamics of a relativistic fluid is presented in terms of the energy-momentum tensor, $\theta^{\mu\nu}$, and the equations of motion are just the conservation equations $\partial_\mu \theta^{\mu\nu} = 0$ [28]. [We denote the relativistic energy-momentum tensor by $\theta^{\mu\nu}$, to distinguish it from the nonrelativistic $T^{\mu\nu}$ introduced in (2.4) and (2.5). The limiting relation between the two is given below.] But we prefer to begin with a Lagrange density

$$\mathcal{L} = -j^\mu a_\mu - f\big(\sqrt{j^\mu j_\mu}\big). \tag{3.61}$$

Here j^μ is the current Lorentz vector $j^\mu = (c\rho, \boldsymbol{j})$. The a_μ comprise a set of auxilliary variables; in the relativistic analog of irrotational fluids we take $a_\mu = \partial_\mu \theta$, more generally,

$$a_\mu = \partial_\mu \theta + \alpha\, \partial_\mu \beta, \tag{3.62}$$

so that the Chern–Simons density of a_i is a total derivative [compare (2.56), (2.57)]. The function f depends on the Lorentz invariant $j^\mu j_\mu = c^2\rho^2 - \boldsymbol{j}^2$ and encodes the specific dyanamics (equation of state).

The energy-momentum tensor for $\mathcal{L}$ is

$$\theta_{\mu\nu} = -g_{\mu\nu}\mathcal{L} + \frac{j_\mu j_\nu}{\sqrt{j^\alpha j_\alpha}} f'\big(\sqrt{j^\alpha j_\alpha}\big). \tag{3.63}$$

[One way to derive (3.63) from (3.61) is to embed that expression in an external metric tensor $g_{\mu\nu}$, which is then varied; in the variation, j^μ and a_μ are taken to be metric independent and $j_\mu = g_{\mu\nu} j^\nu$.] Furthermore, varying j^μ in (3.61) shows that

$$a_\mu = -\frac{j_\mu}{\sqrt{j^\alpha j_\alpha}} f'\big(\sqrt{j^\alpha j_\alpha}\big), \tag{3.64}$$

so that (3.63) becomes

$$\theta_{\mu\nu} = -g_{\mu\nu}[nf'(n) - f(n)] + u_\mu u_\nu n f'(n), \tag{3.65}$$

where we have introduced the proper velocity u_μ by

$$j_\mu = n u_\mu, \qquad u^\mu u_\mu = 1, \tag{3.66}$$

so that n is proportional to the proper density and $1/n$ is proportional to the specific volume. Equation (3.65) identifies the proper energy density e and the pressure P (which coincides with $\mathcal{L}$) through the conventional formula [28]

$$\theta_{\mu\nu} = -g_{\mu\nu} P + u_\mu u_\nu (P + e). \tag{3.67}$$

Therefore, in our case,

$$e = f(n), \tag{3.68}$$
$$P = n f'(n) - f(n). \tag{3.69}$$

The thermodynamic relation involving entropy S reads

$$P\, d\left(\frac{1}{n}\right) + d\left(\frac{e}{n}\right) \propto dS, \tag{3.70}$$

where the proportionality constant is determined by the temperature. With (3.68) and (3.69), the left side of (3.70) vanishes and we verify that entropy is constant, that is, we are dealing with an isentropic system, as has been stated from the very beginning.

For the free system, the pressure vanishes, so we choose $f(n) = cn$:

$$\mathcal{L}_0 = -j^\mu a_\mu - c\sqrt{j^\mu j_\mu}. \tag{3.71}$$

In the "irrotational" case, $a_\mu = \partial_\mu \theta$, and with $\boldsymbol{j} = \rho \boldsymbol{v}$, this Lagrange density produces the free $\bar{L}_0^{\text{Lorentz}}$ of (3.38), (3.39) (apart from a total time derivative).

For the Born–Infeld theory, we present the pressure P in (3.69) by choosing $f(n) = c\sqrt{a^2 + n^2}$, which corresponds to the pressure $P = -a^2 c/\sqrt{a^2 + n^2}$. When the current is written as

$$j_\mu = \frac{a\, \partial_\mu \theta}{\sqrt{c^2 - (\partial_\mu \theta)^2}}, \tag{3.72a}$$

so that

$$n = a\sqrt{\frac{(\partial_\mu \theta)^2}{c^2 - (\partial_\mu \theta)^2}}, \tag{3.72b}$$

the Lagrange density

$$\mathcal{L} = P = -\frac{a^2 c}{\sqrt{a^2 + n^2}}, \tag{3.72c}$$

coincides with that for the Born–Infeld model in (3.46).

Other forms for f give rise to relativistic fluid mechanics with other equations of state.

It is interesting to see how the nonrelativistic limit of $\theta^{\mu\nu}$ in (3.65) produces $T^{\mu\nu}$ of (2.4)–(2.5). It is especially intriguing to notice that $\theta^{\mu\nu}$ is symmetric but $T^{\mu\nu}$ is not. To make the connection we recall that $u^\mu = 1/\sqrt{1 - v^2/c^2}(1, \boldsymbol{v}/c)$, we observe that $n = \sqrt{\rho^2 c^2 - \boldsymbol{j}^2}$, set $\boldsymbol{j} = \rho \boldsymbol{v}$, and conclude that $n = \rho c\sqrt{1 - v^2/c^2} \sim \rho c - (\rho v^2/2c)$. Also $f(n)$ is chosen to be $cn + V(n/c)$, and thus $P = nf'(n) - f(n) = (n/c)V'(n/c) - V(n/c)$. It follows that

$$\begin{aligned}
\theta^{00} &= \frac{nc - (v^2 n/c^3)V'}{1 - v^2/c^2} + V \approx \frac{\rho c^2 - \rho v^2/2}{1 - v^2/c^2} + V(\rho) \\
&\approx \rho c^2 + \frac{\rho v^2}{2} + V(\rho) = \rho c^2 + T^{00}.
\end{aligned} \tag{3.73}$$

Thus, apart from the relativistic "rest energy" ρc^2, θ^{00} passes to T^{00}. The relativistic energy flux is $c\theta^{j0}$ (because $\partial \theta^{\mu 0}/\partial x^\mu = c^{-1}\dot{\theta}^{00} + \partial_j \theta^{j0}$)

$$\begin{aligned}
c\theta^{j0} &= \frac{v^j}{1 - v^2/c^2}\left(nc + \frac{n}{c}V'\right) \approx v^j \frac{\rho c^2 - \rho v^2/2 + \rho V'(\rho)}{1 - v^2/c^2} \\
&\approx j^j c^2 + \rho v^j \left(v^2/2 + V'(\rho)\right) = j^j c + T^{j0}.
\end{aligned} \tag{3.74}$$

Again, apart from the $O(c^2)$ current, associated with the $O(c^2)$ rest energy in θ^{00}, T^{j0} is obtained in the limit. The momentum density is θ^{0i}/c (because $\theta^{\mu\nu}$ has dimension of energy density). Thus

$$\theta^{oi}/c = \frac{v^i/c^2}{1 - v^2/c^2}\left(nc + \frac{n}{c}V'\right) \approx \rho v^i = \mathcal{P}^i. \tag{3.75}$$

Finally, the momentum flux is obtained directly from θ^{ij}:

$$\begin{aligned}\theta^{ij} &= \delta^{ij}\left(\frac{n}{c}V' - V\right) + \frac{v^i v^j}{c^2 - v^2}\left(nc + \frac{n}{c}V'\right)\\ &\approx \delta^{ij}\big(\rho V'(\rho) - V(\rho)\big) + v^i v^j \rho = T^{ij}.\end{aligned} \tag{3.76}$$

From the limiting formula $n \sim \rho c$ we also see that the pressure in (3.72c) tends to the Chaplygin expression $-a/\rho$.

4

Common Ancestry: The Nambu–Goto Action

The "hidden" symmetries and the associated transformation laws for the Chaplygin and Born–Infeld models may be given a coherent setting by considering the Nambu–Goto action for a d-brane in $(d+1)$ spatial dimensions, moving on $(d+1,1)$-dimensional space–time. In our context, a d-brane is simply a d-dimensional extended object a 1-brane is a string, a 2-brane is a membrane, and so on. A d-brane in $(d+1)$ space divides that space in two.

The Nambu–Goto action reads

$$I_{\mathrm{NG}} = \int d\varphi^0\, d\varphi\, \mathcal{L}_{\mathrm{NG}} = \int d\varphi^0\, d\varphi^1 \cdots d\varphi^d \sqrt{G}, \tag{4.1}$$

$$G = (-1)^d \det \frac{\partial X^\mu}{\partial \varphi^\alpha} \frac{\partial X_\mu}{\partial \varphi^\beta}. \tag{4.2}$$

Here X^μ is a $(d+1,1)$ target space–time (d-brane) variable, with μ extending over the range $\mu = 0, 1, \ldots, d, d+1$. The φ^α are "world-volume" variables describing the extended object with α ranging from $\alpha = 0, 1, \ldots, d$; φ^α $(\alpha = 1, \ldots, d)$ parametrizes the d-dimensional d-brane that evolves in φ^0.

The Nambu–Goto action is parametrization invariant, and we shall show that two different choices of parametrization ("light-cone" and "Cartesian") lead to the Chaplygin gas and Born–Infeld actions, respectively. For both parametrizations we choose $(X^1, \ldots, X^d)$ to coincide with $(\varphi^1, \ldots, \varphi^d)$, renaming them as $\boldsymbol{r}$ (a d-dimensional vector). This is usually called the "static parametrization." (The ability to carry out this parametrization globally presupposes that the extended object is topologically trivial; in the contrary situation, singularities will appear, which are spurious in the sense that they disappear in different parametrizations, and parametrization-invariant quantities are singularity-free.)

4.1 Light-Cone Parameterization

For the light-cone parametrization we define $X^{\pm}$ as $(X^0 \pm X^{d+1})/\sqrt{2}$. X^+ is renamed t and identified with $\sqrt{2\lambda}\varphi^0$. This completes the fixing of the parametrization and the remaining variable is X^-, which is a function of φ^0 and $\boldsymbol{\varphi}$ or, after redefinitions, of t and $\boldsymbol{r}$. X^- is renamed as $\theta(t, \boldsymbol{r})$ and then the Nambu–Goto action in this parametrization coincides with the Chaplygin gas action $I_\lambda^{\text{Chaplygin}}$ in (3.26) [29].

4.2 Cartesian Parameterization

For the second, Cartesian parametrization X^0 is renamed ct and identified with $c\varphi^0$. The remaining target space variable X^{d+1}, a function of φ^0 and $\boldsymbol{\varphi}$, equivalently of t and $\boldsymbol{r}$, is renamed $\theta(t, \boldsymbol{r})/c$. Then the Nambu–Goto action reduces to the Born–Infeld action $\int dt\, L_a^{\text{Born–Infeld}}$, (3.46) [29].

4.3 Hodographic Transformation

There is another derivation of the Chaplygin gas from the Nambu–Goto action that makes use of a hodographic transformation, in which independent and dependent variables are interchanged. Although the derivation is more involved than the light-cone/static/parametrization used in Section 4.1, the hodographic approach is instructive in that it gives a natural definition for the density ρ which, in the above static parametrization approach, is determined from θ by the Bernoulli equation (3.14).

We again use light-cone combinations: $(X^0 + X^{d+1})/\sqrt{2}$ is called τ and is identified with φ^0, while $(X^0 - X^{d+1})/\sqrt{2}$ is renamed θ. At this stage, the dependent, target-space variables are θ and the transverse coordinates $\boldsymbol{X} : X^i$ $(i = 1, \dots, d)$, and all are functions of the world-volume parameters $\varphi^0 = \tau$ and $\boldsymbol{\varphi} : \varphi^r$ $(r = 1, \dots, d)$; ∂_τ indicates differentiation with respect to $\tau = \varphi^0$, while ∂_r denotes derivatives with respect to φ^r. The induced metric $G_{\alpha\beta} = \partial X^\mu / \partial\varphi^\alpha \, \partial X_\mu / \partial\varphi^\beta$ takes the form

$$\begin{aligned} G_{\alpha\beta} &= \begin{pmatrix} G_{00} & G_{0s} \\ G_{r0} & -g_{rs} \end{pmatrix} \\ &= \begin{pmatrix} 2\partial_\tau\theta - (\partial_\tau \boldsymbol{X})^2 & \partial_s\theta - \partial_\tau \boldsymbol{X} \cdot \partial_s \boldsymbol{X} \\ \partial_r\theta - \partial_r \boldsymbol{X} \cdot \partial_\tau \boldsymbol{X} & -\partial_r \boldsymbol{X} \cdot \partial_s \boldsymbol{X} \end{pmatrix} . \end{aligned} \tag{4.3}$$

The Nambu–Goto Lagrangian now leads to the canonical momenta

$$\frac{\partial \mathcal{L}_{\text{NG}}}{\partial \partial_\tau \boldsymbol{X}} = \boldsymbol{p}, \tag{4.4a}$$

$$\frac{\partial \mathcal{L}_{\text{NG}}}{\partial \partial_\tau \theta} = \Pi, \tag{4.4b}$$

and can be presented in first-order form as

$$\mathcal{L}_{\text{NG}} = \boldsymbol{p} \cdot \partial_\tau \boldsymbol{X} + \Pi \, \partial_\tau \theta + \frac{1}{2\Pi}(p^2 + g) + u^r (\boldsymbol{p} \cdot \partial_r \boldsymbol{X} + \Pi \, \partial_r \theta), \tag{4.5}$$

where $g = \det g_{rs}$ and

$$u_r \equiv \partial_\tau \boldsymbol{X} \cdot \partial_r \boldsymbol{X} - \partial_r \theta, \tag{4.6}$$

acts as a Lagrange multiplier. Evidently the equations of motion are

$$\partial_\tau \boldsymbol{X} = -\frac{1}{\Pi}\boldsymbol{p} - u^r \, \partial_r \boldsymbol{X}, \tag{4.7a}$$

$$\partial_\tau \theta = \frac{1}{2\Pi^2}(p^2 + g) - u^r \, \partial_r \theta, \tag{4.7b}$$

$$\partial_\tau \boldsymbol{p} = -\partial_r \left(\frac{1}{\Pi} g g^{rs} \, \partial_s \boldsymbol{X} \right) - \partial_r (u^r \boldsymbol{p}), \tag{4.7c}$$

$$\partial_\tau \Pi = -\partial_r (u^r \Pi). \tag{4.7d}$$

Also there is the constraint

$$\boldsymbol{p}\cdot\partial_r\boldsymbol{X}+\Pi\,\partial_r\theta=0. \tag{4.8}$$

[That u^r is still given by (4.6) is a consequence of (4.7a) and (4.8).] Here g^{rs} is inverse to g_{rs}, and the two metrics are used to move the (r,s) indices. The theory still possesses an invariance against redefining the spatial parameters with a τ-dependent function of the parameters; infinitesimally: $\delta\varphi^r = -f^r(\tau,\boldsymbol{\varphi})$, $\delta\theta = f^r\,\partial_r\theta$, $\delta X^i = f^r\,\partial_r X^i$. This freedom may be used to set u^r to zero and Π to -1.

Next the hodographic transformation is performed: Rather than viewing the dependent variables $\boldsymbol{p}$, θ, and $\boldsymbol{X}$ as functions of τ and $\boldsymbol{\varphi}$, $\boldsymbol{X}(\tau,\boldsymbol{\varphi})$ is inverted so that $\boldsymbol{\varphi}$ becomes a function of τ and $\boldsymbol{X}$ (renamed t and $\boldsymbol{r}$, respectively), and $\boldsymbol{p}$ and θ also become functions of t and $\boldsymbol{r}$. It then follows from the chain rule that the constraint (4.8) (at $\Pi=-1$) becomes

$$0=\frac{\partial X^i}{\partial\varphi^r}\left(p^i-\frac{\partial}{\partial X^i}\theta\right), \tag{4.9}$$

and is solved by

$$\boldsymbol{p}=\boldsymbol{\nabla}\theta. \tag{4.10}$$

Moreover, according to the chain rule and the implicit function theorem, the partial derivative with respect to τ at fixed $\boldsymbol{\varphi}$ [this derivative is present in (4.5)] is related to the partial derivative with respect to τ at fixed $\boldsymbol{X}=\boldsymbol{r}$ by

$$\partial_\tau=\frac{\partial}{\partial t}+\boldsymbol{\nabla}\theta\cdot\boldsymbol{\nabla}, \tag{4.11}$$

where we have used the new name "t" on the right. Thus the Nambu–Goto Lagrangian—the φ integral of the Lagrange density (4.5) (at $u^r=0$, $\Pi=-1$)—reads

$$L_{\text{NG}}=\int d\varphi\{\boldsymbol{p}\cdot\boldsymbol{\nabla}\theta-\dot\theta-\boldsymbol{\nabla}\theta\cdot\boldsymbol{\nabla}\theta-\tfrac{1}{2}(p^2+g)\}. \tag{4.12a}$$

But use of (4.10) and of the Jacobian relation

$$d\varphi=dr\det\frac{\partial\varphi^s}{\partial X^i}=\frac{dr}{\sqrt{g}}$$

shows that

$$L_{\mathrm{NG}} = \int dr \left\{ -\frac{1}{\sqrt{g}}\dot{\theta} - \frac{1}{2\sqrt{g}}(\boldsymbol{\nabla}\theta)^2 - \tfrac{1}{2}\sqrt{g} \right\}. \tag{4.12b}$$

With the definition

$$\sqrt{g} = \sqrt{2\lambda}/\rho, \tag{4.12c}$$

L_{NG} becomes, apart from a total time derivative

$$L_{\mathrm{NG}} = \frac{1}{\sqrt{2\lambda}} \int dr \left\{ \theta\dot{\rho} - \tfrac{1}{2}\rho(\boldsymbol{\nabla}\theta)^2 - \frac{\lambda}{\rho} \right\}. \tag{4.12d}$$

Up to an overall factor, this is just the Chaplygin gas Lagrangian in (3.13).

The present derivation has the advantage of relating the density ρ to the Jacobian of the $\boldsymbol{X}^i \to \boldsymbol{\varphi}$ transformation: $\rho = \sqrt{2\lambda}\det \partial\varphi^s/\partial X^i$. (This in turn shows that the hodographic transformation is just exactly the passage from Lagrangian to Eulerian fluid variables—a remark aimed at those who are acquainted with the Lagrange formulation of fluid motion [3].)

Finally, let us observe that the expansion of the Galileo symmetry in $(d,1)$ space–time to a Poincaré symmetry in $(d+1,1)$ space–time can be understood from a Kaluza–Klein-type framework [30].

4.4 Interrelations

The relation to the Nambu–Goto action explains the origin of the hidden $(d+1,1)$ Poincaré group in our two nonlinear models on $(d,1)$ space–time: Poincaré invariance is what remains of the reparametrization invariance of the Nambu–Goto action after choosing either the light-cone or Cartesian parametrizations. Also the nonlinear, field-dependent form of the transformation laws leading to the additional symmetries is understood: it arises from the identification of some of the dependent variables (X^μ) with the independent variables (φ^α).

The complete integrability of the $d = 1$ Chaplygin gas and Born–Infeld model is a consequence of the fact that both descend from a string in 2-space; the associated Nambu–Goto theory being completely integrable. We shall discuss this in Chapter 6.

We observe that in addition to the nonrelativistic descent from the Born–Infeld theory to the Chaplygin gas, there exists a mapping of one system on another, and between solutions of one system and the other, because both have the same d-brane ancestor. The mapping is achieved by passing from the light-cone parametrization to the Cartesian, or vice versa. Specifically, this is accomplished as follows.

Chaplygin Gas → Born–Infeld

Given $\theta_{\mathrm{NR}}(t, \boldsymbol{r})$, a nonrelativistic solution, determine $T(t, \boldsymbol{r})$ from the equation

$$T + \frac{1}{c^2}\theta_{\mathrm{NR}}(T, \boldsymbol{r}) = \sqrt{2}t. \tag{4.13}$$

Then the relativistic solution is

$$\theta_R(t, \boldsymbol{r}) = \frac{1}{\sqrt{2}}c^2T - \frac{1}{\sqrt{2}}\theta_{\mathrm{NR}}(T, \boldsymbol{r}) = c^2(\sqrt{2}T - t). \tag{4.14}$$

Born–Infeld → Chaplygin Gas

Given $\theta_R(t, \boldsymbol{r})$, a relativistic solution, find $T(t, \boldsymbol{r})$ from

$$T + \frac{1}{c^2}\theta_R(T, \boldsymbol{r}) = \sqrt{2}t. \tag{4.15}$$

Then the nonrelativistic solution is

$$\theta_{\mathrm{NR}}(t, \boldsymbol{r}) = \frac{1}{\sqrt{2}}c^2T - \frac{1}{\sqrt{2}}\theta_R(T, \boldsymbol{r}) = c^2(\sqrt{2}T - t). \tag{4.16}$$

The relation between the different models is depicted in Figure 4.1.

As a final comment, we recall that the elimination of ρ, both in the nonrelativistic (Chaplygin) and relativistic (Born–Infeld)

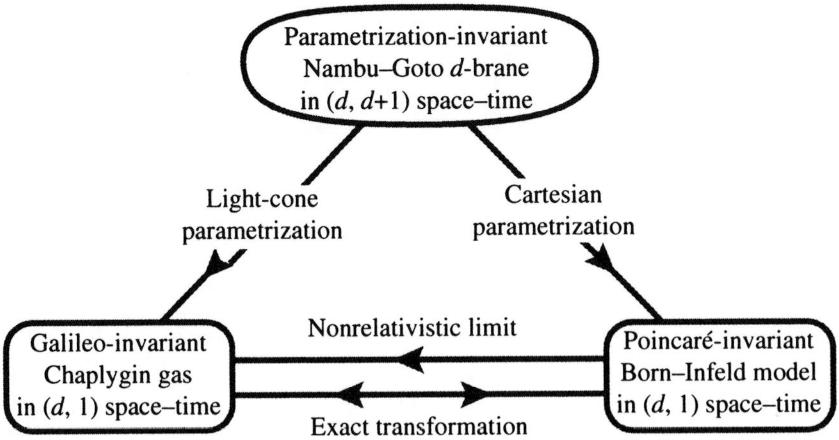

FIGURE 4.1. Dualities and other relations between nonlinear equations.

models is possible only in the presence of interactions. Nevertheless, the θ-dependent (ρ-independent) resultant Lagrangians contain the interaction strengths only as overall factors; see (3.26) and (3.46). It is these θ-dependent Lagrangians that correspond to the Nambu–Goto action in various parametrizations. Let us further recall that the Nambu–Goto action also carries an overall multiplicative factor: the d-brane "tension," which has been suppressed in (4.1). Correspondingly, for a "tensionless" d-brane, the Nambu–Goto expression vanishes, and cannot generate dynamics. This suggests that an action for "tensionless" d-branes could be the noninteracting fluid mechanical expressions (3.13), (3.44), with vanishing coupling strengths λ and a, respectively. Furthermore, we recall that the noninteracting models retain the higher, dynamical symmetries, appropriate to a d-brane in one higher dimension.

5

Supersymmetric Generalization

Once proven the fact that a bosonic Nambu–Goto theory gives rise to and links together the Chaplygin gas and Born–Infeld models, which are irrotational in that the velocity of the former and the momentum of the latter are given by a gradient of a potential, one can ask whether there is a d-brane that produces a fluid model with nonvanishing vorticity.

We shall show that this indeed can be achieved if one starts with a super-d-brane and, moreover, the resulting fluid model possesses supersymmetry. However, since theories of extended "super" objects cannot be formulated in arbitrary dimensions, we shall consider the fluid in two spatial dimensions, namely, on the plane [31].

5.1 Chaplygin Gas with Grassmann Variables

We begin by positing the fluid model. The Chaplygin gas Lagrangian in (3.13) is supplemented by Grassmann variables ψ_a that are Majorana spinors [real, two-component: $\psi_a^* = \psi_a$, $a = 1, 2$, $(\psi_1\psi_2)^* = \psi_1^*\psi_2^*$].

The associated Lagrange density reads

$$\mathcal{L} = -\rho(\dot{\theta} - \tfrac{1}{2}\psi\dot{\psi}) - \tfrac{1}{2}\rho(\boldsymbol{\nabla}\theta - \tfrac{1}{2}\psi\boldsymbol{\nabla}\psi)^2 - \frac{\lambda}{\rho} - \frac{\sqrt{2\lambda}}{2}\psi\boldsymbol{\alpha}\cdot\boldsymbol{\nabla}\psi. \quad (5.1)$$

Here α^i are two ($i = 1, 2$), 2×2, real symmetric Dirac "alpha" matrices; in terms of Pauli matrices we can take $\alpha^1 = \sigma^1$, $\alpha^2 = \sigma^3$. Note that the matrices satisfy the following relations, which are needed to verify subsequent formulas:

$$\begin{aligned}\varepsilon_{ab}\alpha^i_{bc} &= \varepsilon^{ij}\alpha^j_{ac},\\ \alpha^i_{ab}\alpha^j_{bc} &= \delta^{ij}\delta_{ac} - \varepsilon^{ij}\varepsilon_{ac},\\ \alpha^i_{ab}\alpha^i_{cd} &= \delta_{ac}\delta_{bd} - \delta_{ab}\delta_{cd} + \delta_{ad}\delta_{bc},\end{aligned} \tag{5.2}$$

where ε_{ab} is the 2×2 antisymmetric matrix $\varepsilon \equiv i\sigma^2$. In (5.1) λ is a coupling strength which is assumed to be positive. The Grassmann term enters with coupling $\sqrt{2\lambda}$, which is correlated with the strength of the Chaplygin potential $V(\rho) = \lambda/\rho$ in order to ensure supersymmetry, as we shall show below. It is evident that the velocity should be defined as

$$\boldsymbol{v} = \boldsymbol{\nabla}\theta - \tfrac{1}{2}\psi\boldsymbol{\nabla}\psi. \tag{5.3}$$

The Grassmann variables directly give rise to a Clebsch formula for $\boldsymbol{v}$, and provide the Gauss potentials. The two-dimensional vorticity reads $\omega = \varepsilon^{ij}\,\partial_i v^j = -\frac{1}{2}\varepsilon^{ij}\,\partial_i\psi\,\partial_j\psi = -\frac{1}{2}\boldsymbol{\nabla}\psi \times \boldsymbol{\nabla}\psi$. The variables $\{\theta, \rho\}$ remain a canonical pair, while the canonical 1-form in (5.1) indicates that the canonically independent Grassmann variables are $\sqrt{\rho}\,\psi$ so that the antibracket of the ψ's is

$$\{\psi_a(\boldsymbol{r}), \psi_b(\boldsymbol{r}')\} = -\frac{\delta_{ab}}{\rho(\boldsymbol{r})}\delta(\boldsymbol{r} - \boldsymbol{r}'). \tag{5.4}$$

One verifies that the algebra (2.14) or (2.16) is satisfied, and further, one has

$$\{\theta(\boldsymbol{r}), \psi(\boldsymbol{r})\} = -\frac{1}{2\rho(\boldsymbol{r})}\psi(\boldsymbol{r})\delta(\boldsymbol{r} - \boldsymbol{r}'), \tag{5.5}$$

$$\{\boldsymbol{v}(\boldsymbol{r}), \psi(\boldsymbol{r}')\} = -\frac{\boldsymbol{\nabla}\psi(\boldsymbol{r})}{\rho(\boldsymbol{r})}\delta(\boldsymbol{r} - \boldsymbol{r}'), \tag{5.6}$$

$$\{\boldsymbol{\mathcal{P}}(\boldsymbol{r}), \psi(\boldsymbol{r}')\} = -\boldsymbol{\nabla}\psi(\boldsymbol{r})\delta(\boldsymbol{r} - \boldsymbol{r}'). \tag{5.7}$$

The momentum density $\boldsymbol{\mathcal{P}}$ is given by the bosonic formula $\boldsymbol{\mathcal{P}} = \rho\boldsymbol{v}$, but the Grassmann variables are hidden in $\boldsymbol{v}$, by virtue of (5.3).

The equations of motion read

$$\dot{\rho} + \boldsymbol{\nabla} \cdot (\rho \boldsymbol{v}) = 0, \tag{5.8}$$

$$\dot{\theta} + \boldsymbol{v} \cdot \boldsymbol{\nabla}\theta = \tfrac{1}{2}v^2 + \frac{\lambda}{\rho^2} + \frac{\sqrt{2\lambda}}{2\rho}\psi \boldsymbol{\alpha} \cdot \boldsymbol{\nabla}\psi, \tag{5.9}$$

$$\dot{\psi} + \boldsymbol{v} \cdot \boldsymbol{\nabla}\psi = \frac{\sqrt{2\lambda}}{\rho}\boldsymbol{\alpha} \cdot \boldsymbol{\nabla}\psi, \tag{5.10}$$

and together with (5.3) they imply

$$\dot{\boldsymbol{v}} + \boldsymbol{v} \cdot \boldsymbol{\nabla}\boldsymbol{v} = \boldsymbol{\nabla}\frac{\lambda}{\rho^2} + \frac{\sqrt{2\lambda}}{\rho}(\boldsymbol{\nabla}\psi)\boldsymbol{\alpha} \cdot \boldsymbol{\nabla}\psi. \tag{5.11}$$

All these equations may be obtained by bracketing with the Hamiltonian

$$H = \int d^2r \left(\tfrac{1}{2}\rho v^2 + \frac{\lambda}{\rho} + \frac{\sqrt{2\lambda}}{2}\psi \boldsymbol{\alpha} \cdot \boldsymbol{\nabla}\psi \right) = \int d^2\, r\mathcal{H}, \tag{5.12}$$

when (2.14), (2.16), as well as (5.4)–(5.6), are used. We record the components of the energy-momentum "tensor," and the continuity equations they satisfy. The energy density $\mathcal{E} = T^{00}$, given by

$$\mathcal{E} = \tfrac{1}{2}\rho v^2 + \frac{\lambda}{\rho} + \frac{\sqrt{2\lambda}}{2}\psi \boldsymbol{\alpha} \cdot \boldsymbol{\nabla}\psi = T^{00}, \tag{5.13}$$

satisfies a continuity equation with the energy flux T^{j0}:

$$\begin{aligned} T^{j0} = \rho v^j \left(\tfrac{1}{2}v^2 - \frac{\lambda}{\rho^2} \right) &+ \frac{\sqrt{2\lambda}}{2}\psi \alpha^j \boldsymbol{v} \cdot \boldsymbol{\nabla}\psi \\ &- \frac{\lambda}{\rho}\psi \partial_j \psi + \frac{\lambda}{\rho}\varepsilon^{jk}\psi \varepsilon \partial_k \psi, \end{aligned} \tag{5.14}$$

$$\dot{T}^{00} + \partial_j T^{j0} = 0. \tag{5.15}$$

This ensures that the total energy, that is, the Hamiltonian, is time-independent. Conservation of the total momentum,

$$\boldsymbol{P} = \int d^2r\, \boldsymbol{\mathcal{P}}, \tag{5.16}$$

follows from the continuity equation satisfied by the momentum density $\mathcal{P}^i = T^{0i}$ and the momentum flux, that is, the stress tensor T^{ij}:

$$T^{ji} = \rho v^i v^j - \delta^{ij}\left(\frac{2\lambda}{\rho} + \frac{\sqrt{2\lambda}}{2}\psi\boldsymbol{\alpha}\cdot\boldsymbol{\nabla}\psi\right) + \frac{\sqrt{2\lambda}}{2}\psi\alpha^j\,\partial_i\psi, \tag{5.17}$$

$$\dot{T}^{0i} + \partial_j T^{ji} = 0. \tag{5.18}$$

But T^{ij} is not symmetric in its spatial indices, owing to the presence of spin in the problem. However, rotational symmetry makes it possible to effect an "improvement," which modifies the momentum density by a total derivative term, leaving the integrated total momentum unchanged (provided surface terms can be ignored) and rendering the stress tensor symmetric. The improved quantities are

$$\mathcal{P}_I^i = T_I^{0i} = \rho v^i + \tfrac{1}{8}\varepsilon^{ij}\partial_j(\rho\psi\varepsilon\psi) \tag{5.19}$$

$$T_I^{ij} = \rho v^i v^j - \delta^{ij}\left(\frac{2\lambda}{\rho} + \frac{\sqrt{2\lambda}}{2}\psi\boldsymbol{\alpha}\cdot\boldsymbol{\nabla}\psi\right) + \frac{\sqrt{2\lambda}}{4}(\psi\alpha^i\,\partial_j\psi + \psi\alpha^j\,\partial_i\psi)$$
$$- \tfrac{1}{8}\partial_k[(\varepsilon^{ki}v^j + \varepsilon^{kj}v^i)\rho\psi\varepsilon\psi], \tag{5.20}$$

$$\dot{T}_I^{0i} + \partial_j T_I^{ij} = 0. \tag{5.21}$$

It immediately follows from the symmetry of T_I^{ij} that the angular momentum,

$$M = \int d^2r\,\varepsilon^{ij}r^i\mathcal{P}_I^j = \int d^2r\,\rho\varepsilon^{ij}r^iv^j + \frac{1}{4}\int d^2r\,\rho\psi\varepsilon\psi \tag{5.22}$$

is conserved. The first term is clearly the orbital part (which still receives a Grassmann contribution through $\boldsymbol{v}$), whereas the second, coming from the improvement, is the spin part. Indeed, since $(i/2)\varepsilon = \frac{1}{2}\sigma^2 \equiv \Sigma$, we recognize this as the spin matrix in $(2+1)$ dimensions. The extra term in the improved momentum density (5.19), $\frac{1}{8}\varepsilon^{ij}\partial_j(\rho\psi\varepsilon\psi)$, can then be readily interpreted as an additional localized momentum density, generated by the nonhomogeneity of the spin density. This is analogous to the magnetostatics formula giving the localized current density $\boldsymbol{j}_m$

in a magnet in terms of its magnetization $\boldsymbol{m}$: $\boldsymbol{j}_m = \boldsymbol{\nabla} \times \boldsymbol{m}$. All in all, we are describing a fluid with spin.

Also the total number

$$N = \int d^2r\, \rho, \tag{5.23}$$

is conserved by virtue of the continuity equation (5.8) satisfied by ρ. Finally, the theory is Galileo invariant, as is seen from the conservation of the Galileo boost,

$$\boldsymbol{B} = t\boldsymbol{P} - \int d^2r\, \boldsymbol{r}\rho \tag{5.24}$$

which follows from (5.8) and (5.16). The generators $H, \boldsymbol{P}, M, \boldsymbol{B}$, and N close on the (extended) Galileo group. [The theory is not Lorentz invariant in (2 + 1)-dimensional space–time, hence the energy flux T^{j0} does not coincide with the momentum density, improved or not.]

We observe that ρ can be eliminated from (5.1) so that $\mathcal{L}$ involves only θ and ψ. From (5.9) and (5.10) it follows that

$$\rho = \sqrt{\lambda}(\dot{\theta} - \tfrac{1}{2}\psi\dot{\psi} + \tfrac{1}{2}v^2)^{-1/2}. \tag{5.25}$$

Substituting into (5.1) produces the supersymmetric generalization of the Chaplygin gas Lagrange density in (3.26)

$$\mathcal{L} = -\,2\sqrt{\lambda}\big\{\sqrt{2\dot{\theta} - \psi\dot{\psi} + (\boldsymbol{\nabla}\theta - \tfrac{1}{2}\psi\boldsymbol{\nabla}\psi)^2} + \tfrac{1}{2}\psi\boldsymbol{\alpha}\cdot\boldsymbol{\nabla}\psi\big\}. \tag{5.26}$$

Note that the coupling strength has disappeared from the dynamical equations, remaining only as a normalization factor for the Lagrangian. Consequently, the above elimination of ρ cannot be carried out in the free case, $\lambda = 0$.

5.2 Supersymmetry

As we said earlier, this theory possesses supersymmetry. This can be established, first of all, by verifying that the following two-component supercharges are time-independent Grassmann quantities

$$Q_a = \int d^2r\,[\rho\boldsymbol{v}\cdot(\boldsymbol{\alpha}_{ab}\psi_b) + \sqrt{2\lambda}\psi_a]. \tag{5.27}$$

Taking a time derivative and using the evolution equations (5.8)–(5.11) establishes that $\dot{Q}_a = 0$.

Next, the supersymmetric transformation rule for the dynamical variables is found by constructing a bosonic symmetry generator Q, obtained by contracting the Grassmann charge with a constant Grassmann parameter η^a, $Q = \eta^a Q_a$, and commuting with the dynamical variables. Using the canonical brackets one verifies the following field transformation rules:

$$\delta\rho = \{Q, \rho\} = -\boldsymbol{\nabla}\cdot\rho(\eta\boldsymbol{\alpha}\psi), \tag{5.28}$$

$$\delta\theta = \{Q, \theta\} = -\tfrac{1}{2}(\eta\boldsymbol{\alpha}\psi)\cdot\boldsymbol{\nabla}\theta - \tfrac{1}{4}(\eta\boldsymbol{\alpha}\psi)\cdot\psi\boldsymbol{\nabla}\psi + \frac{\sqrt{2\lambda}}{2\rho}\eta\psi, \tag{5.29}$$

$$\delta\psi = \{Q, \psi\} = -(\eta\boldsymbol{\alpha}\psi)\cdot\boldsymbol{\nabla}\psi - \boldsymbol{v}\cdot\boldsymbol{\alpha}\eta - \frac{\sqrt{2\lambda}}{\rho}\eta, \tag{5.30}$$

$$\delta\boldsymbol{v} = \{Q, \boldsymbol{v}\} = -(\eta\boldsymbol{\alpha}\psi)\cdot\boldsymbol{\nabla}\boldsymbol{v} + \frac{\sqrt{2\lambda}}{\rho}\eta\boldsymbol{\nabla}\psi. \tag{5.31}$$

Supersymmetry is reestablished by determining the variation of the action $\int dt\, d^2r\, \mathcal{L}$ consequent to the above field variations: the action is invariant. One then reconstructs the supercharges (5.27) by Noether's theorem. Finally, upon computing the bracket of two supercharges, one finds

$$\{\eta_1^a Q_a, \eta_2^b Q_b\} = 2(\eta_1\eta_2)H, \tag{5.32}$$

which again confirms that the charges are time independent:

$$\{H, Q_a\} = 0. \tag{5.33}$$

Additionally, a further, kinematical supersymmetry can be identified. According to the equations of motion the following two supercharges are also time independent:

$$\overline{Q}_a = \int d^2r\,\rho\psi_a. \tag{5.34}$$

$\overline{Q} = \bar{\eta}^a \overline{Q}_a$ effects a shift of the Grassmann field:

$$\bar{\delta}\rho = \{\overline{Q}, \rho\} = 0, \tag{5.35}$$
$$\bar{\delta}\theta = \{\overline{Q}, \theta\} = -\tfrac{1}{2}(\bar{\eta}\psi), \tag{5.36}$$
$$\bar{\delta}\psi = \{\overline{Q}, \psi\} = -\bar{\eta}, \tag{5.37}$$
$$\bar{\delta}\boldsymbol{v} = \{\overline{Q}, \boldsymbol{v}\} = 0. \tag{5.38}$$

This transformation leaves the Lagrangian invariant, and Noether's theorem reproduces (5.34). The algebra of these charges closes on the total number N:

$$\{\bar{\eta}_1^a \overline{Q}_a, \bar{\eta}_2^b \overline{Q}_b\} = (\bar{\eta}_1 \bar{\eta}_2) N, \tag{5.39}$$

while the algebra with the generators (5.27) closes on the total momentum, together with a central extension, proportional to volume of space $\Omega = \int d^2 r$:

$$\{\bar{\eta}^a \overline{Q}_a, \eta^b Q_b\} = (\bar{\eta}\boldsymbol{\alpha}\eta) \cdot \boldsymbol{P} + \sqrt{2\lambda}\,(\bar{\eta}\eta)\Omega. \tag{5.40}$$

The supercharges Q_a, $\overline{Q}_a$, together with the Galileo generators $(H, \boldsymbol{P}, M$, and $\boldsymbol{B})$, with N form a superextended Galileo algebra. The additional, nonvanishing brackets are

$$\{M, Q_a\} = \tfrac{1}{2}\varepsilon^{ab} Q_b, \tag{5.41}$$
$$\{M, \overline{Q}_a\} = \tfrac{1}{2}\varepsilon^{ab} \overline{Q}_b, \tag{5.42}$$
$$\{\boldsymbol{B}, Q_a\} = \boldsymbol{\alpha}_{ab} \overline{Q}_b. \tag{5.43}$$

5.3 Supermembrane Connection

The equations for the supersymmetric Chaplygin fluid devolve from a supermembrane Lagrangian, L_M. We shall give two different derivations of this result, which make use of two different parametrizations for the parametrization-invariant membrane action and give rise, respectively, to (5.1) and (5.26). These two derivations follow what has been done in the bosonic case in Sections 4.1 and 4.3.

We work in a light-cone gauge-fixed theory: The supermembrane in four-dimensional space–time is described by coordinates X^μ ($\mu = 0, 1, 2, 3$), which are decomposed into light-cone components $X^\pm = (X^0 \pm X^3)/\sqrt{2}$ and transverse components X^i $\{i = 1, 2\}$. These depend on an evolution parameter $\varphi^0 \equiv \tau$ and two space-like parameters φ^r $\{r = 1, 2\}$. Additionally, there are two-component, real Grassmann spinors ψ, which also depend on τ and φ^r. In the light-cone gauge, X^+ is identified with τ, X^- is renamed θ, and the supermembrane Lagrangian is [32]

$$L_M = \int d^2\varphi\, \mathcal{L}_M = -\int d^2\varphi \{\sqrt{G} - \tfrac{1}{2}\varepsilon^{rs}\, \partial_r\psi\boldsymbol{\alpha}\, \partial_s\psi \cdot \boldsymbol{X}\}, \tag{5.44}$$

where $G = \det G_{\alpha\beta}$;

$$\begin{aligned} G_{\alpha\beta} &= \begin{pmatrix} G_{00} & G_{0s} \\ G_{r0} & -g_{rs} \end{pmatrix} \\ &= \begin{pmatrix} 2\partial_\tau\theta - (\partial_\tau \boldsymbol{X})^2 - \psi\partial_\tau\psi & u_s \\ u_r & -g_{rs} \end{pmatrix}, \end{aligned} \tag{5.45}$$

$$\begin{aligned} G &= g\Gamma, \\ \Gamma &\equiv 2\,\partial_\tau\theta - (\partial_\tau \boldsymbol{X})^2 - \psi\,\partial_\tau\psi + g^{rs}u_r u_s, \\ g_{rs} &\equiv \partial_r \boldsymbol{X} \cdot \partial_s \boldsymbol{X}, \qquad g = \det g_{rs}, \\ u_s &\equiv \partial_s\theta - \tfrac{1}{2}\psi\,\partial_s\psi - \partial_\tau \boldsymbol{X} \cdot \partial_s \boldsymbol{X}. \end{aligned} \tag{5.46}$$

Here ∂_τ signifies differentiation with respect to the evolution parameter τ, while ∂_r differentiates with respect to the space-like parameters φ^r; g^{rs} is the inverse of g_{rs}, and the two are used to move the (r, s) indices. Note that the dimensionality of the transverse coordinates X^i is the same as that of the parameters φ^r, namely two.

5.4 Hodographic Transformation

To give our first derivation following the procedure in Section 4.3, we rewrite the Lagrangian in canonical, first-order form, with the help of bosonic canonical momenta defined by

$$\frac{\partial \mathcal{L}_M}{\partial \partial_\tau \boldsymbol{X}} = \boldsymbol{p} = -\Pi\, \partial_\tau \boldsymbol{X} - \Pi u^r\, \partial_r \boldsymbol{X}, \tag{5.47a}$$

$$\frac{\partial \mathcal{L}_M}{\partial \partial_\tau \theta} = \Pi = \sqrt{g/\Gamma}. \tag{5.47b}$$

(The Grassmann variables already enter with first-order derivatives.) The supersymmetric extension of (4.5) then reads

$$\begin{aligned}\mathcal{L}_M = \boldsymbol{p}\cdot\partial_\tau \boldsymbol{X} + \Pi\,\partial_\tau\theta - \tfrac{1}{2}\Pi\psi\,\partial_\tau\psi \\ + \frac{1}{2\Pi}(p^2+g) + \tfrac{1}{2}\varepsilon^{rs}\,\partial_r\psi\boldsymbol{\alpha}\,\partial_s\psi\cdot\boldsymbol{X} \\ + u^r(\boldsymbol{p}\cdot\partial_r\boldsymbol{X} + \Pi\,\partial_r\theta - \tfrac{1}{2}\Pi\psi\,\partial_r\psi).\end{aligned} \tag{5.48}$$

In (5.48) u^r serves as a Lagrange multiplier enforcing a subsidiary condition on the canonical variables, and $g = \det g_{rs}$. The equations that follow from (5.48) coincide with the Euler–Lagrange equations for (5.44). The theory still possesses an invariance against redefining the spatial parameters with a τ-dependent function of the parameters. This freedom may be used to set u_τ to zero and fix Π at -1. Next we introduce the hodographic transformation, as in Section 4.3, whereby independent-dependent variables are interchanged, namely, we view the φ^r to be functions of X^i. It then follows that the constraint on (5.48) which, with $\Pi = -1$, reads

$$\boldsymbol{p}\cdot\partial_r\boldsymbol{X} - \partial_r\theta + \tfrac{1}{2}\psi\,\partial_r\psi = 0, \tag{5.49}$$

becomes

$$\partial_r\boldsymbol{X}\cdot(\boldsymbol{p} - \boldsymbol{\nabla}\theta + \tfrac{1}{2}\psi\boldsymbol{\nabla}\psi) = 0. \tag{5.50}$$

Here $\boldsymbol{p}$, θ, and ψ are viewed as functions of $\boldsymbol{X}$, renamed $\boldsymbol{r}$, with respect to which acts the gradient $\boldsymbol{\nabla}$. Also, we rename $\boldsymbol{p}$ as $\boldsymbol{v}$ which, according to (5.50), is

$$\boldsymbol{v} = \boldsymbol{\nabla}\theta - \tfrac{1}{2}\psi\boldsymbol{\nabla}\psi. \tag{5.51}$$

As in Section 4.3, from the chain rule and the implicit function theorem it follows that

$$\partial_\tau = \partial_t + \partial_\tau\boldsymbol{X}\cdot\boldsymbol{\nabla}, \tag{5.52}$$

and according to (5.47a) (at $\Pi = -1$, $u^r = 0$) $\partial_\tau \boldsymbol{X} = \boldsymbol{p} = \boldsymbol{v}$. Finally, the measure transforms according to $d^2\varphi \to d^2r/\sqrt{g}$. Thus the Lagrangian for (5.48) becomes, after setting u^r to zero and Π to -1,

$$L_M = \int \frac{d^2r}{\sqrt{g}} \Big(v^2 - \dot\theta - \boldsymbol{v}\cdot\boldsymbol{\nabla}\theta + \tfrac{1}{2}\psi(\dot\psi + \boldsymbol{v}\cdot\boldsymbol{\nabla}\psi) \\ - \tfrac{1}{2}(v^2+g) - \tfrac{1}{2}\varepsilon^{rs}\psi\alpha^i\,\partial_j\psi\,\partial_s x^j\,\partial_r x^i \Big). \tag{5.53a}$$

But $\varepsilon^{rs}\,\partial_s x^j\,\partial_r x^i = \varepsilon^{ij}\det\partial_r x^i = \varepsilon^{ij}\sqrt{g}$. After $\sqrt{g}$ is renamed $\sqrt{2\lambda}/\rho$, (5.53a) finally reads

$$L_M = \frac{1}{\sqrt{2\lambda}} \int d^2r \Big(-\rho(\dot\theta - \tfrac{1}{2}\psi\dot\psi) - \tfrac{1}{2}\rho(\boldsymbol{\nabla}\theta - \tfrac{1}{2}\psi\boldsymbol{\nabla}\psi)^2 \\ - \frac{\lambda}{\rho} - \frac{\sqrt{2\lambda}}{2}\psi\boldsymbol{\alpha}\times\boldsymbol{\nabla}\psi \Big). \tag{5.53b}$$

Upon replacing ψ by $(1-\varepsilon)\psi/\sqrt{2}$, this is seen to reproduce the Lagrange density (5.1), apart from an overall factor.

5.5 Light-Cone Parametrization

For our second derivation, we return to (5.44)–(5.46) and use the remaining reparametrization freedom to equate the two X^i variables with the two φ^r variables, renaming both as r^i. Also τ is renamed as t. This parallels the method in Section 4.1. Now, in (5.44)–(5.46), $g_{rs} = \delta_{rs}$ and $\partial_\tau \boldsymbol{X} = 0$, so that (5.46) becomes simply

$$G = \Gamma = 2\dot\theta - \psi\dot\psi + u^2, \tag{5.54}$$

$$\boldsymbol{u} = \boldsymbol{\nabla}\theta - \tfrac{1}{2}\psi\boldsymbol{\nabla}\psi. \tag{5.55}$$

Therefore the supermembrane Lagrangian (5.44) reads

$$L_M = -\int d^2r \{\sqrt{2\dot\theta - \psi\dot\psi + (\boldsymbol{\nabla}\theta - \tfrac{1}{2}\psi\boldsymbol{\nabla}\psi)^2} + \tfrac{1}{2}\psi\boldsymbol{\alpha}\times\boldsymbol{\nabla}\psi\}. \tag{5.56}$$

Again a replacement of ψ by $(1-\varepsilon)\psi/\sqrt{2}$ demonstrates that the integrand coincides with the Lagrange density in (5.26) (apart from a normalization factor).

5.6 Further Consequences of the Supermembrane Connection

Supermembrane dynamics is Poincaré invariant in (3 + 1)-dimensional space–time. This invariance is hidden by the choice of light-cone parametrization: only the light-cone subgroup of the Poincaré group is left as a manifest invariance. This is just the (2+1) Galileo group generated by $H, \boldsymbol{P}, M, \boldsymbol{B}$, and N. (The light-cone subgroup of the Poincaré group is isomorphic to the Galileo group in one lower dimension [33].) The Poincaré generators not included in the above list correspond to Lorentz transformations in the "–" direction. We expect therefore that these generators are "dynamical," that is, hidden and unexpected conserved quantities of our supersymmetric Chaplygin gas, similar to the situation with the purely bosonic model.

One verifies that the following quantities

$$D = tH - \int d^2r\, \rho\theta, \tag{5.57}$$

$$\begin{aligned} \boldsymbol{G} &= \int d^2r(\boldsymbol{r}\mathcal{H} - \theta\boldsymbol{\mathcal{P}}_I - \tfrac{1}{8}\psi\alpha\boldsymbol{\alpha}\cdot\boldsymbol{\mathcal{P}}_I\psi) \\ &= \int d^2r(\boldsymbol{r}\mathcal{H} - \theta\boldsymbol{\mathcal{P}} - \tfrac{1}{4}\psi\alpha\boldsymbol{\alpha}\cdot\boldsymbol{\mathcal{P}}\psi), \end{aligned} \tag{5.58}$$

are time independent by virtue of the equations of motion (5.8)–(5.11), and they supplement the Galileo generators to form the full (3+1) Poincaré algebra, which becomes the super-Poincaré algebra once the supersymmetry is taken into account. Evidently, (5.57), (5.58) are the supersymmetric generalizations of (3.23), (3.24).

We see that fluid dynamics can be extended to include Grassmann variables, which also enter into a supersymmetry-preserving interaction. Since our construction is based on a supermembrane in (3+1)-dimensional space–time, the fluid model is necessarily a planar Chaplygin gas. It remains for the future to show how this construction could be generalized to arbitrary dimensions and to different interactions. Note that Grassmann–Gauss potentials ψ can be used even in the absence of supersymmetry. For example, our theory (5.1), with the last term omitted, possesses a conventional, bosonic Hamiltonian without supersymmetry, while the Grassmann variables are hidden in $\boldsymbol{v}$ and occur only in the canonical 1-form.

6

One-Dimensional Case

In this section, we shall discuss the nonrelativistic–relativistic models in one spatial dimension. Complete integrability has been established for both the Chaplygin gas [34] and the Born–Infeld theory [35]. We can now understand this to be a consequence of the complete integrability of the Nambu–Goto 1-brane (string) moving on a 2-space (plane), which is the antecedent of both models. [Therefore, it suffices to discuss only the Chaplygin gas since solutions of the Born–Infeld model can then be obtained by the mapping (4.13), (4.14).]

As remarked previously, in one dimension there is no vorticity, and the nonrelativistic velocity v can be presented as a derivative with respect to the single spatial variable of a potential θ. Similarly, the relativistic momentum $p = v/\sqrt{1 - v^2/c^2}$ is a derivative of a potential θ. In both cases the potential is canonically conjugate to the density ρ governed by the canonical 1-form $\int dx\, \theta \dot{\rho}$. Moreover, it is evident that at the expense of a spatial nonlocality, one may replace θ by its antiderivative, which is p both nonrelativistically and relativistically (nonrelativistically $p = v$), so that in both cases the Lagrangian reads

$$L = -\frac{1}{2} \int dx\, dy\, \rho(x) \varepsilon(x - y) \dot{p}(y) - H. \tag{6.1}$$

For the Chaplygin gas and the Born–Infeld models, H is given, respectively, by

$$H^{\text{Chaplygin}} = \int dx \left(\tfrac{1}{2} \rho p^2 + \frac{\lambda}{\rho} \right), \tag{6.2}$$

$$H^{\text{Born–Infeld}} = \int dx \left(\sqrt{\rho^2 c^2 + a^2} \sqrt{c^2 + p^2} \right). \tag{6.3}$$

The equations of motion are, respectively,

Chaplygin gas:

$$\dot{\rho} + \frac{\partial}{\partial x}(p\rho) = 0, \tag{6.4}$$

$$\dot{p} + \frac{\partial}{\partial x}\left(\frac{p^2}{2} - \frac{\lambda}{\rho^2}\right) = 0, \tag{6.5}$$

or

$$\frac{\partial}{\partial t}\frac{1}{\sqrt{\dot{\theta} + p^2/2}} + \frac{\partial}{\partial x}\frac{p}{\sqrt{\dot{\theta} + p^2/2}} = 0. \tag{6.6}$$

Born–Infeld model:

$$\dot{\rho} + \frac{\partial}{\partial x}\left(p\sqrt{\frac{\rho^2c^2 + a^2}{c^2 + p^2}}\right) = 0, \tag{6.7}$$

$$\dot{p} + \frac{\partial}{\partial x}\left(\rho c^2\sqrt{\frac{c^2 + p^2}{\rho^2c^2 + a^2}}\right) = 0, \tag{6.8}$$

or

$$\frac{\partial}{c^2\partial t}\left(\frac{\dot{\theta}}{\sqrt{c^2 - c^{-2}\dot{\theta}^2 + p^2}}\right) - \frac{\partial}{\partial x}\left(\frac{p}{\sqrt{c^2 - c^{-2}\dot{\theta}^2 + p^2}}\right) = 0. \tag{6.9}$$

In the above, (6.6) and (6.9) result by determining ρ in terms of θ ($p = \partial\theta/\partial x$) from (6.5) and (6.8), and using that expression for ρ in (6.4) and (6.7).

6.1 Specific Solutions for the Chaplygin Gas on a Line

Classes of solutions for a Chaplygin gas in one dimension can be given in closed form. For example, to obtain general, time-rescaling-invariant solutions, we make the *Ansatz* that $\theta \propto 1/t$. Then (3.27), or (6.6), leads to a second-order nonlinear differential equation for the x-dependence of θ. Therefore, solutions involve two arbitrary constants, one of which fixes the origin of x (we suppress it); the other we call k, and take it to be real. The solutions then read

$$\theta(t, x) = -\frac{1}{2k^2t}\cosh^2 kx. \tag{6.10}$$

[Other solutions can be obtained by relaxing the reality condition on k and/or shifting the argument kx by a complex number. In this way one finds that θ can also be $\sinh^2 kx/(2k^2 t)$, $\sin^2 kx/(2k^2 t)$, $\cos^2 kx/(2k^2 t)$; but these lead to singular or unphysical forms for ρ.] The density corresponding to (6.10) is found from (3.14) or (6.5) to be

$$\rho(t,x) = \sqrt{2\lambda}\frac{k|t|}{\cosh^2 kx}. \tag{6.11}$$

The velocity/momentum $v = p = \partial\theta/\partial x$ is

$$v(t,x) = p(t,x) = -\frac{1}{kt}\sinh kx \cosh kx, \tag{6.12}$$

while the sound speed

$$s(t,x) = \frac{\cosh^2 kx}{k|t|}, \tag{6.13}$$

is always larger than $|v|$. Finally, the current $j = \rho\,\partial\theta/\partial x$ exhibits a kink profile,

$$j(t,x) = -\varepsilon(t)\sqrt{2\lambda}\tanh kx, \tag{6.14}$$

which is suggestive of complete integrability.

Another particular solution is the Galileo boost of the static profiles (3.32), (3.33):

$$p(t,x) = p(x-ut), \tag{6.15}$$

$$\rho(t,x) = \frac{\sqrt{2\lambda}}{|p-u|}. \tag{6.16}$$

Here u is the boosting velocity and $p(x-ut)$ is an arbitrary function of its argument (provided $p \neq u$). Clearly this is a constant profile solution, in linear motion with velocity u.

Further evidence for complete integrability is found by identifying an infinite number of constants of motion. One verifies that the following quantities

$$I_n^{\pm} = \int dx\,\rho\left(p \pm \frac{\sqrt{2\lambda}}{\rho}\right)^n, \qquad n = 0, \pm 1, \ldots, \tag{6.17}$$

are conserved.

The combinations $p \pm \sqrt{2\lambda}/\rho$ are just the velocity ± the sound speed, and they are known as Riemann coordinates

$$R_{\pm} = p \pm \frac{\sqrt{2\lambda}}{\rho}. \tag{6.18}$$

The equations of motion for this system [continuity (6.4) and Euler (6.5)] can be succinctly presented in terms of $R_{\pm}$:

$$\dot{R}_{\pm} = -R_{\mp}\frac{\partial}{\partial x}R_{\pm}. \tag{6.19}$$

6.2 Aside on the Integrability of the Cubic Potential in One Dimension

Although it does not belong to the models that we have discussed, the cubic potential for one-dimensional motion, $V(\rho) = \ell\rho^3/3$, is especially interesting because it is secretly free—a fact that is exposed when Riemann coordinates are employed. For this problem these read $R_{\pm} = p \pm \sqrt{2\ell}\rho$ and again they are just the velocity ± the sound speed. In contrast to (6.19) the Euler and continuity equations for this system decouple: $\dot{R}_{\pm} = -R_{\pm}\partial/\partial x R_{\pm}$. Indeed, it is seen that $R_{\pm}$ satisfy essentially the free Euler equation [compare with (2.42) and identify $R_{\pm}$ with v]. Consequently, the solution (2.44)–(2.46) works here as well.

Recall the previous remark in Section 3.1 on the Schrödinger group [Galileo ⊕ SO(2, 1)]: in one dimension the cubic potential is invariant against this group of transformations, and in all dimensions the free theory is invariant [17, 18]. Therefore, a natural speculation is that the secretly noninteracting nature of the cubic potential in one dimension is a consequence of Schrödinger group invariance.

Another interesting fact about a one-dimensional nonrelativistic fluid with cubic potential is that it also arises in a collective, semiclassical description of nonrelativistic free fermions in one dimension, where the cubic potential reproduces fermion repulsion [36]. In spite of the nonlinearity of the fluid model's equations of motion, there is no interaction in the underlying fermion dynamics. Thus, the presence of the Schrödinger group and the equivalence to free equations for this fluid system is an understandable consequence.

6.3 General Solution for the Chaplygin Gas on a Line

The general solution to the Chaplygin gas can be found by linearizing the governing equations (continuity and Euler) with the help of a Legendre transform, which also effects a hodographic transformation that exchanges the independent variables (t, x) with the dependent ones (ρ, θ); actually, instead of ρ we use the sound speed $s = \sqrt{2\lambda}/\rho$ and instead of θ we use the momentum $p = \partial\theta/\partial x$.

Define

$$\psi(p, s) = \theta(t, x) - t\dot{\theta}(t, x) - x\frac{\partial}{\partial x}\theta(t, x). \tag{6.20}$$

From the Bernoulli equation we know that

$$\dot{\theta} = -\tfrac{1}{2}p^2 + \tfrac{1}{2}s^2. \tag{6.21}$$

Thus

$$\psi(p, s) = \theta(t, x) + \frac{t}{2}(p^2 - s^2) - xp, \tag{6.22}$$

and the usual Legendre transform rules govern the derivatives

$$\frac{\partial\psi}{\partial p} = tp - x, \tag{6.23a}$$

$$\frac{\partial\psi}{\partial s} = -ts. \tag{6.23b}$$

It remains to incorporate the continuity equation (6.4) whose content must be recast by the hodographic transformation. This is achieved by rewriting equation (6.4) in terms of $s = \sqrt{2\lambda}/\rho$:

$$\frac{\partial s}{\partial t} + p\frac{\partial s}{\partial x} - s\frac{\partial p}{\partial x} = 0. \tag{6.24}$$

Next, (6.24) is presented as a relation between Jacobians

$$\frac{\partial(s,x)}{\partial(t,x)} + p\frac{\partial(t,s)}{\partial(t,x)} - s\frac{\partial(t,p)}{\partial(t,x)} = 0, \tag{6.25a}$$

which is true because here $\partial x/\partial t = \partial t/\partial x = 0$. Equation (6.25a) implies, after multiplication by $\partial(t,x)/\partial(s,p)$,

$$\begin{aligned} 0 &= \frac{\partial(s,x)}{\partial(s,p)} + p\frac{\partial(t,s)}{\partial(s,p)} - s\frac{\partial(t,p)}{\partial(s,p)} \\ &= \frac{\partial x}{\partial p} - p\frac{\partial t}{\partial p} - s\frac{\partial t}{\partial s}. \end{aligned} \tag{6.25b}$$

The second equality holds because now we take $\partial s/\partial p = \partial p/\partial s = 0$. Finally, from (6.22), (6.23) it follows that (6.25b) is equivalent to

$$\frac{\partial^2 \psi}{\partial p^2} - \frac{\partial^2 \psi}{\partial s^2} + \frac{2}{s}\frac{\partial \psi}{\partial s} = 0. \tag{6.25c}$$

This linear equation is solved by two arbitrary functions of $p \pm s$ ($p \pm s$ being just the Riemann coordinates)

$$\psi(p,s) = F(p+s) - sF'(p+s) + G(p-s) + sG'(p-s). \tag{6.26}$$

In summary, to solve the Chaplygin gas equations, we choose two functions F and G, construct ψ as in (6.26), and regain s ($= \sqrt{2\lambda}/\rho$), p ($= \partial\theta/\partial x$), and θ from (6.22), (6.23). In particular, the solution (6.10), (6.11) corresponds to

$$F(z) = G(-z) = \pm\frac{z}{2k}\ln z, \tag{6.27}$$

where the sign is correlated with the sign of t.

6.4 Born–Infeld Model on a Line

Since the Born–Infeld system is related to Chaplygin gas by the transformation described in Section 4.4, there is no need to discuss, separately, Born–Infeld solutions. Nevertheless, the formulation in terms of Riemann coordinates is especially succinct and gives another view on the Chaplygin/Born–Infeld relation.

The Riemann coordinates $R_\pm$ for the Born–Infeld model are constructed by first defining

$$\begin{aligned} \frac{1}{c}\frac{\partial}{\partial x}\theta = p/c = \tan\varphi_p, \\ a/\rho c = \tan\varphi_\rho, \end{aligned} \tag{6.28}$$

and

$$R_\pm = \varphi_p \pm \varphi_\rho. \tag{6.29}$$

The one-dimensional version of the equations of motion (3.47), (3.48), that is, (6.7), (6.8), can be presented as

$$\dot{R}_\pm = -c(\sin R_\mp)\frac{\partial}{\partial x}R_\pm. \tag{6.30}$$

The relation to the Riemann description of the Chaplygin gas can now be seen in two ways: a nonrelativistic limit and an exact transformation. For the former, we note that at large c, $\varphi_p \approx p/c$, $\varphi_\rho \approx a/\rho c$ so that

$$R_\pm^{\text{Born–Infeld}} \approx \frac{1}{c}\left(p \pm \frac{a}{\rho}\right) = \frac{1}{c}R_\pm^{\text{Chaplygin}}\Big|_{\lambda=a^2/2}. \tag{6.31}$$

Moreover, equation (6.30) becomes, in view of (6.31),

$$\frac{1}{c}\dot{R}_\pm^{\text{Chaplygin}} = -R_\mp^{\text{Chaplygin}}\frac{1}{c}\frac{\partial}{\partial x}R_\pm^{\text{Chaplygin}}, \tag{6.32}$$

so that (6.19) is regained. On the other hand, for the exact transformation we define new Riemann coordinates in the relativistic, Born–Infeld case by

$$\mathcal{R}_\pm = c\sin R_\pm. \tag{6.33}$$

Evidently, (6.30) implies that $\mathcal{R}_\pm$ satisfies the nonrelativistic equations (6.19), (6.32) when $R_\pm$ solves the relativistic equation (6.30). Expressing $\mathcal{R}_\pm$ and $R_\pm$ in terms of the corresponding nonrelativistic and relativistic variables produces a mapping between the two sets. Calling p_{NR}, ρ_{NR} and p_{R}, ρ_{R} the momentum and density of the nonrelativistic and relativistic theory, respectively, the mapping implied by (6.33) is

$$\begin{aligned} p_{\mathrm{NR}} &= \frac{c^2 \rho_{\mathrm{R}} p_{\mathrm{R}}}{\sqrt{(p_{\mathrm{R}}^2 + c^2)(\rho_{\mathrm{R}}^2 c^2 + a^2)}}, \\ \rho_{\mathrm{NR}} &= \frac{1}{c^2}\sqrt{(p_{\mathrm{R}}^2 + c^2)(\rho_{\mathrm{R}}^2 c^2 + a^2)}. \end{aligned} \tag{6.34}$$

As can be checked, this maps the Chaplygin equations into the Born–Infeld equations. But the mapping is not canonical.

We record the infinite number of constants of motion, which put into evidence the (by now obvious) complete integrability of the Born–Infeld equations on a line. The following quantities are time independent:

$$I_n^{\pm} = a c^{n-1} \int dx \, \frac{(\varphi_p \pm \varphi_\rho)^n}{\sin \varphi_\rho \cos \varphi_p}, \qquad n = 0, \pm 1, \ldots. \tag{6.35}$$

The nonrelativistic limit takes the above into (6.17), while expressing $I_n^{\pm}$ in terms of $\mathcal{R}_{\pm}$ according to (6.33) shows that the integrals in (6.35) are expressible as series in terms of the integrals in (6.17).

In the relativistic model, ρ need not be constrained to be positive (negative ρ could be interpreted as antiparticle density). The transformation $p \to -p$, $\rho \to -\rho$ is a symmetry and can be interpreted as charge conjugation. Further, p and ρ appear in an equivalent way. As a result, this theory enjoys a duality transformation

$$\rho \to \pm \frac{a}{c^2} p, \qquad p \to \pm \frac{c^2}{a} \rho. \tag{6.36}$$

Under the above, both the canonical structure and the Hamiltonian remain invariant. Solutions are mapped in general to new solutions. Note that the nonrelativistic limit is mapped to the ultrarelativistic limit under the above duality. Self-dual solutions, with $\rho = \pm a/c^2 p$, satisfy

$$\dot{\rho} = \mp c \frac{\partial}{\partial x} \rho, \tag{6.37}$$

and are, therefore, the chiral relativistic solutions that were presented at the end of Section 3.2. In the self-dual case, when p is eliminated from the canonical 1-form, and from the Hamiltonian with the help of (6.36), one arrives at an action for ρ, which coincides (apart from irrelevant constants) with the self-dual action, constructed some time ago [37]

$$\begin{aligned}
&\left\{\frac{1}{2}\int dt\,dx\,dy\,\dot{\rho}(x)\varepsilon(x-y)p(y)\right. \\
&\qquad\qquad \left.-\int dt\,dx\,\sqrt{\rho^2c^2+a^2}\sqrt{c^2+p^2}\,dt\right\}\Bigg|_{p=c^2\rho/a} \\
&\quad=\frac{2c^2}{a}\left\{\frac{1}{4}\int dt\,dx\,dy\,\dot{\rho}(x)\varepsilon(x-y)\rho(y)\right. \\
&\qquad\qquad \left.-\frac{c}{2}\int dt\,dx\left(\rho^2(x)+\frac{a^2}{c^2}\right)\right\}. \qquad (6.38)
\end{aligned}$$

6.5 General Solution of the Nambu–Goto Theory for a ($d = 1$)-Brane (String) in Two Spatial Dimensions (on a Plane)

The complete integrability of the Chaplygin gas and of the Born–Infeld theory, as well as the relationships between the two, derives from the fact that the different models descend by fixing, in different ways, the parameterization invariance of the Nambu–Goto theory for a string on a plane. At the same time, the equations governing the planar motion of a string can be solved completely. Therefore, it is instructive to see how the string solution produces this Chaplygin solution [21].

We follow the development in Section 4.3. The Nambu–Goto action reads

$$I_{\mathrm{NG}} = \int d\varphi^0\, L_{\mathrm{NG}}, \qquad (6.39\mathrm{a})$$

$$L_{\mathrm{NG}} = \int d\varphi^1\, \mathcal{L}_{\mathrm{NG}}, \qquad (6.39\mathrm{b})$$

$$\mathcal{L}_{\mathrm{NG}} = \left[-\det\frac{\partial X^\mu}{\partial\varphi^\alpha}\frac{\partial X_\mu}{\partial\varphi^\beta}\right]^{1/2}. \qquad (6.39\mathrm{c})$$

Here X^μ ($\mu = 0, 1, 2$) are string variables and (φ^0, φ^1) are its parameters. As in Section 4.3, we define light-cone combinations $X^\pm = (1/\sqrt{2})(X^0 \pm X^2)$, rename X^- as θ, and choose the parameterization $X^+ = \varphi^0 \equiv \tau$. After suppressing the superscripts on φ^0 and X^1, we construct the Nambu–Goto Lagrange density as

$$\mathcal{L}_{\rm NG} = \overset{1/2}{\det}\begin{pmatrix} 2\partial_\tau\theta - (\partial_\tau X)^2 & u \\ u & -(\partial_\varphi X)^2 \end{pmatrix}, \tag{6.40}$$

$$u = \partial_\varphi\theta - \partial_\tau X\,\partial_\varphi X. \tag{6.41}$$

Equations of motion are presented in Hamiltonian form:

$$p \equiv \frac{\partial\mathcal{L}_{\rm NG}}{\partial\partial_\tau X}, \qquad \Pi \equiv \frac{\partial\mathcal{L}_{\rm NG}}{\partial\partial_\tau\theta}, \tag{6.42}$$

$$\partial_\tau X = -\frac{1}{\Pi}p - u\partial_\varphi X, \tag{6.43a}$$

$$\partial_\tau\theta = \frac{1}{2\Pi^2}(p^2 + (\partial_\varphi X)^2) - u\,\partial_\varphi\theta, \tag{6.43b}$$

$$\partial_\tau p = -\partial_\varphi\left(\frac{1}{\Pi}\partial_\varphi X\right) - \partial_\varphi(up), \tag{6.43c}$$

$$\partial_\tau\Pi = -\partial_\varphi(u\Pi), \tag{6.43d}$$

and there is the constraint

$$p\,\partial_\varphi X + \Pi\,\partial_\varphi\theta = 0. \tag{6.44}$$

There still remains the reparameterization freedom of replacing φ by an arbitrary function of τ and φ; this freedom may be used to set $u = 0$, $\Pi = -1$. Consequently, in the fully parameterized equations of motion (6.43d) disappears; instead of (6.43a) and (6.43c), we have $\partial_\tau X = p$, $\partial_\tau p = \partial_\varphi^2 X$, which imply

$$(\partial_\tau^2 - \partial_\varphi^2)X = 0. \tag{6.45a}$$

Equation (6.43b) reduces to

$$\partial_\tau\theta = \tfrac{1}{2}[(\partial_\tau X)^2 + (\partial_\varphi X)^2], \tag{6.45b}$$

and the constraint (6.44) requires

$$\partial_\varphi \theta = \partial_\tau X \partial_\varphi X. \tag{6.45c}$$

Solution to (6.45a) is immediate in terms of two functions $F_\pm$:

$$x(\tau,\varphi) = F_+(\tau+\varphi) + F_-(\tau-\varphi), \tag{6.46}$$

and then (6.45b), (6.45c) fix θ:

$$\theta(\tau,\varphi) = \int^{\tau+\varphi} dz\,[F'_+(z)]^2 + \int^{\tau-\varphi} dz\,[F'_-(z)]^2. \tag{6.47}$$

This completes the description of a string moving on a plane. But we need to convert this information into a solution of the Chaplygin gas, and we know from Section 4.3 that this can be accomplished by a hodographic transformation: instead of X and θ as a function of τ and φ, we seek φ as a function of τ and X, and this renders θ to be a function of τ and X as well. The density ρ is determined by the Jacobian $|\partial X/\partial\varphi|$.

Replace τ by t and X by x and define φ to be $f(t,x)$. Then from (6.46) it follows that

$$x = F_+\big(t+f(t,x)\big) + F_-\big(t-f(t,x)\big). \tag{6.48}$$

This equation may be differentiated with respect to t and x, whereupon one finds

$$\frac{\partial f}{\partial t} = -\frac{F'_+(t+f)+F'_-(t-f)}{F'_+(t+f)-F'_-(t-f)}, \tag{6.49a}$$

$$\frac{\partial f}{\partial x} = \frac{1}{F'_+(t+f)-F'_-(t-f)}. \tag{6.49b}$$

Thus the procedure for constructing a Chaplygin gas solution is to choose two functions $F_\pm$, solve the differential equations (6.49) for f, and then the fluid variables are

$$\theta(t,x) = \int^{t+f(t,x)} [F'_+(z)]^2\,dz + \int^{t-f(t,x)} [F'_-(z)]^2\,dz, \tag{6.50}$$

$$\frac{\sqrt{2\lambda}}{\rho} = |F'_+(t+f) - F'_-(t-f)|. \tag{6.51}$$

One may verify directly that (6.50) and (6.51) satisfy the required equations: Upon differentiating (6.50) with respect to t and x, we find

$$\begin{aligned}\frac{\partial\theta}{\partial t} &= (F_+')^2\left(1+\frac{\partial f}{\partial t}\right) + (F_-')^2\left(1-\frac{\partial f}{\partial t}\right)\\ &= -2F_+'F_-',\end{aligned} \tag{6.52a}$$

$$\begin{aligned}\frac{\partial\theta}{\partial x} &= (F_+')^2\left(\frac{\partial f}{\partial x}\right) - (F_-')^2\left(\frac{\partial f}{\partial x}\right)\\ &= F_+' + F_-'.\end{aligned} \tag{6.52b}$$

The second equalities follow with the help of (6.49). From (6.52) one sees that

$$\frac{\partial\theta}{\partial t} + \frac{1}{2}\left(\frac{\partial\theta}{\partial x}\right)^2 = \tfrac{1}{2}(F_+' - F_-')^2 = \frac{\lambda}{\rho^2}, \tag{6.53}$$

the last equality being the definition (6.51). Thus the Bernoulli (Euler) equation holds. For the continuity equation, we first find, from (6.51) and (6.52),

$$\begin{aligned}\frac{\partial\rho}{\partial t} &= \pm\frac{\partial}{\partial t}\frac{\sqrt{2\lambda}}{F_+' - F_-'}\\ &= \mp\frac{\sqrt{2\lambda}}{(F_+' - F_-')^2}\left[F_+''\left(1+\frac{\partial f}{\partial t}\right) - F_-''\left(1-\frac{\partial f}{\partial t}\right)\right]\\ &= \pm\frac{2\sqrt{2\lambda}}{(F_+' - F_-')^3}(F_+''F_-' + F_-''F_+'),\end{aligned} \tag{6.54a}$$

$$\begin{aligned}\frac{\partial}{\partial x}\left(\rho\frac{\partial\theta}{\partial x}\right) &= \frac{\partial}{\partial x}\left(\pm\sqrt{2\lambda}\frac{F_+' + F_-'}{F_+' - F_-'}\right)\\ &= \mp\frac{\sqrt{2\lambda}}{(F_+' - F_-')^2}(F_+''F_-' + F_-''F_+')\frac{\partial f}{\partial x}\\ &= \mp\frac{2\sqrt{2\lambda}}{(F_+' - F_-')^3}(F_+''F_-' + F_-''F_+').\end{aligned} \tag{6.54b}$$

The last equalities follow from (6.49); since (6.54a) and (6.54b) sum to zero, the continuity equation holds.

We observe that the differentiated functions $F'_\pm$ are just the Riemann coordinates: from (6.52b) and (6.51) [with the absolute value ignored] we have

$$p \pm \frac{\sqrt{2\lambda}}{\rho} \equiv R_\pm = 2F'_\pm . \tag{6.55}$$

Also it is seen with the help of (6.49) that the Riemann formulation (6.19) of the Chaplygin equations is satisfied by $2F'_\pm$.

The constants of motion (6.17) become proportional to

$$\begin{aligned} I_n^\pm &\propto \int dx \, \frac{1}{F'_+ - F'_-} [F'_\pm]^n \\ &= \int dx \, \frac{\partial f}{\partial x} \left[F'_\pm(t \pm f)\right]^n \\ &\propto \int dz \left[F'_\pm(z)\right]^n . \end{aligned} \tag{6.56}$$

Finally, we remark that the solution (6.10), (6.11) corresponds to

$$F_+(z) = -F_-(z) = \pm \frac{\ln z}{2k} . \tag{6.57}$$

There exists a relation between the two functions F and G in (6.26), which encode the Chaplygin gas solution in the linearization approach of Section 6.3, and the above two functions $F_\pm$, which do the same job in the Nambu–Goto approach. The relation is that $2F'_+$ is inverse to $2F''$, and $2F'_-$ is inverse to $2G''$, that is,

$$\begin{aligned} 2F''[2F'_+(z)] &= z, \\ 2G''[2F'_-(z)] &= z. \end{aligned} \tag{6.58}$$

Problem 6.1. Derive (6.58). Verify this relation with (6.27) and (6.57).

6.6 Supersymmetric Chaplygin Gas on a Line

The planar Chaplygin gas, enhanced with Grassmann variables and supporting a supersymmetry, also exists in one spatial dimension [38]. It descends from the supersymmetrized Nambu–Goto action for a 1-brane on (2 + 1)-dimensional space–time, namely, a superstring on a plane. We record here the emergent Lagrangian

$$L = \frac{1}{\sqrt{2\lambda}} \int dx \left\{ -\rho(\dot\theta - \tfrac{1}{2}\psi\dot\psi) - \tfrac{1}{2}\rho(\partial_x\theta - \tfrac{1}{2}\psi\partial_x\psi)^2 - \frac{\lambda}{\rho} + \frac{\sqrt{2\lambda}}{2}\psi\partial_x\psi \right\} \tag{6.59a}$$

$$L' = -\int dx \sqrt{2\dot\theta - \psi\dot\psi + (\partial_x\theta - \tfrac{1}{2}\psi\partial_x\psi)^2} - \tfrac{1}{2}\psi\partial_x\psi. \tag{6.59b}$$

The second formula follows from the first after ρ is eliminated, and it is L' that emerges from the superstring reduction. ψ is a real Grassmann variable, and the combination $\partial_x\theta - \frac{1}{2}\psi\partial_x\psi$ is the velocity v. [Some sign conventions and normalizations differ from what is found in Chapter 5.]

The equations of motion that follow from varying (6.59a) read

$$\dot\rho + \partial_x(\rho v) = 0, \tag{6.60}$$

$$\dot\psi + \left(v + \frac{\sqrt{2\lambda}}{\rho}\right)\partial_x\psi = 0, \tag{6.61}$$

$$\dot\theta + v\,\partial_x\theta = \tfrac{1}{2}v^2 + \frac{\lambda}{\rho^2} - \frac{\sqrt{2\lambda}}{2\rho}\psi\,\partial_x\psi, \tag{6.62}$$

and they imply

$$\dot v + v\,\partial_x v = \partial_x\left(\frac{\lambda}{\rho^2}\right). \tag{6.63}$$

Note that the continuity (6.60) and Euler (6.63) equations do not see the Grassmann variables directly; hence their solution is the same as in the bosonic case, and this is brought out by expressing these two equations in terms of Riemann coordinates (velocity $\pm$ sound speed $\sqrt{2\lambda}/\rho$) [compare (6.18) and (6.19), where p is used to represent v]

$$R_\pm = \left(v \pm \frac{\sqrt{2\lambda}}{\rho}\right), \tag{6.64}$$

$$\dot R_\pm = -R_\mp\partial_x R_\pm. \tag{6.65}$$

The remaining equations, (6.61) and (6.62), are also expressed in terms of the Riemann coordinates

$$\dot{\psi} = -R_+ \, \partial_x \psi, \tag{6.66}$$
$$\dot{\theta} = -\tfrac{1}{2} R_+ R_- - \tfrac{1}{2} R_+ \psi \, \partial_x \psi. \tag{6.67}$$

The complete integrability of the bosonic theory means that (6.65) are integrable. Given a solution, it follows that (6.66) is solved by any function of R_-:

$$\psi = \Psi(R_-). \tag{6.68}$$

Of course, the complete integrability of the supersymmetric model on a line comes as no surprise, considering that it descends from an integrable world-sheet theory.

It may seem curiously asymmetric that (6.66) for the Grassmann field should contain the R_+ Riemann coordinate and not the R_- companion coordinate. In fact, the reverse would have been the case if the sign of the $\sqrt{2\lambda}$ term in (6.59a) had been opposite. The entire model is consistent with this substitution.

The energy-momentum tensor is constructed from (6.59a), and its components are

$$T^{00} = \mathcal{H} = \tfrac{1}{2}\rho v^2 + \frac{\lambda}{\rho} - \frac{\sqrt{2\lambda}}{2} \psi \, \partial_x \psi, \tag{6.69}$$
$$T^{01} = \mathcal{P} = \rho v, \tag{6.70}$$
$$T^{10} = \frac{\rho v}{2} R_+ R_- - \frac{\sqrt{2\lambda}}{2} R_+ \psi \, \partial_x \psi, \tag{6.71}$$
$$T^{11} = \rho R_+ R_-. \tag{6.72}$$

The expected conserved quantities of the system, the generators of the Galileo group, are verified to be time independent using the equations of motion. We have

$$N = \int dx\ \rho, \tag{6.73}$$

$$P = \int dx\ \rho v, \tag{6.74}$$

$$H = \int dx \left(\tfrac{1}{2}\rho v^2 + \frac{\lambda}{\rho} - \frac{\sqrt{2\lambda}}{2}\psi\, \partial_x \psi \right), \tag{6.75}$$

$$B = \int dx\ \rho(x - vt) = \int dx\ x\rho - tP. \tag{6.76}$$

Although some generators look purely bosonic, there are still Grassmann fields hidden in v.

In going to Riemann coordinates, we can observe a ladder of conserved charges of the form [compare (6.17)]

$$I_n^{\pm} = \int dx\ \rho R_{\pm}^n. \tag{6.77}$$

The first few values of n above give

$$I_0^{\pm} = N, \tag{6.78}$$

$$I_1^{\pm} = P \pm \sqrt{2\lambda}\Omega, \tag{6.79}$$

$$I_2^{+} = 2H, \tag{6.80}$$

where Ω is used to denote the length of space $\int dx$. (Note that I_2^- would correspond to the Hamiltonian of the theory with $\sqrt{2\lambda}$ replaced by its negative.)

In Section 5.2 there are identified two different supersymmetry generators for the planar model. These correspond in one space dimension to the time-independent quantities

$$\tilde{Q} = \int dx\ \rho\psi, \tag{6.81}$$

$$Q = \int dx\ \rho \left(v - \frac{\sqrt{2\lambda}}{\rho} \right) \psi. \tag{6.82}$$

These, again, are but special cases ($n = 0$ and $n = 1$) of a ladder of conserved supercharges described by

$$Q_n = \int dx \; \rho R_-^n \psi. \tag{6.83}$$

We see that the supercharges evaluated on the solution (6.68) reproduce the form of the bosonic charges (6.77).

Let us observe that there exist further bosonic and fermionic conserved charges. For example, one may verify that the bosonic charges

$$\int dx \; \rho R_\pm^n \left(\frac{\partial_x R_\pm}{\rho} \right)^m, \tag{6.84}$$

$$\int dx \; \rho R_-^n \left(\frac{\psi \partial_x \psi}{\rho} \right), \tag{6.85}$$

are conserved, as are the fermionic charges

$$\int dx \; \rho R_-^n \left(\frac{\partial_x \psi}{\rho} \right). \tag{6.86}$$

Conserved expressions involving higher derivatives may also be constructed. The conservation of these quantities is easily understood when the string world-sheet variables are used. Then the above are written as $\int d\sigma \; R_\pm^n (\partial_\sigma R_\pm)^m$, $\int d\sigma \; R_-^n (\psi \partial_\sigma \psi)$, and $\int d\sigma \; R_-^n (\partial_\sigma \psi)$, respectively. Furthermore, when $R_\pm$ are evaluated on solutions, they become functions of $\tau \pm \sigma$ (see Section 6.5), so that integration over σ extinguishes the τ dependence, leaving constant quantities.

The equations of motion (6.60)–(6.62) can also be obtained by Poisson bracketing with the Hamiltonian (6.75) if the following canonical brackets are postulated:

$$\{\theta(x), \rho(y)\} = \delta(x - y), \tag{6.87}$$

$$\{\theta(x), \psi(y)\} = -\frac{\psi}{2\rho} \delta(x - y), \tag{6.88}$$

$$\{\psi(x), \psi(y)\} = -\frac{1}{\rho} \delta(x - y), \tag{6.89}$$

where the last bracket, containing Grassmann arguments, is understood to be the antibracket. With these one verifies that the conserved charges in (6.73)–(6.76) generate the appropriate Galileo symmetry transformations on the dynamical variables ρ, θ, and ψ. Correspondingly, the supercharges (6.81), (6.82) generate the supertransformations

$$\tilde{\delta}\rho = 0, \qquad \delta\rho = -\eta\,\partial_x(\rho\psi), \tag{6.90}$$

$$\tilde{\delta}\theta = -\tfrac{1}{2}\eta\psi, \qquad \delta\theta = -\tfrac{1}{2}\eta R_+\psi, \tag{6.91}$$

$$\tilde{\delta}\psi = -\eta, \qquad \delta\psi = -\eta\psi\partial_x\psi - \eta R_-, \tag{6.92}$$

which leave the Lagrangian (6.59a) invariant. The algebra of the bosonic generators reproduces the algebra of the (extended) Galileo group, the extension residing in the bracket $\{B, P\} = -N$. The algebra of the supercharges is

$$\{\bar{\eta}Q, \eta Q\} = 2\bar{\eta}\eta H, \tag{6.93}$$

$$\{\bar{\eta}\tilde{Q}, \eta\tilde{Q}\} = \bar{\eta}\eta N, \tag{6.94}$$

$$\{\bar{\eta}\tilde{Q}, \eta Q\} = \bar{\eta}\eta(P - \sqrt{2\lambda}\Omega), \tag{6.95}$$

$$\{B, Q\} = \tilde{Q}. \tag{6.96}$$

Since the fluid model descends from the superstring, it should possess an enhanced symmetry beyond the Galileo symmetry in (1 + 1) dimensions. In fact, the following conserved charges, affecting time rescaling and space–time mixing, are also verified [compare (5.57), (5.58)]:

$$D = \int dx\,(t\mathcal{H} - \rho\theta), \tag{6.97}$$

$$G = \int dx\,(x\mathcal{H} - \theta\mathcal{P}). \tag{6.98}$$

The Galileo generators supplemented by D and G together, satisfy the Lie algebra of the (2+1)-dimensional Poincaré group, with N, P, and H corresponding to the three translations and with B, D, and G forming the (2+1)-dimensional Lorentz group SO(2, 1):

$$\{B, D\} = B, \qquad \{G, B\} = D, \qquad \{D, G\} = G, \tag{6.99}$$

with Casimir

$$C = B \circ G + G \circ B + D \circ D. \tag{6.100}$$

Adjoining the supercharges results in the super-Poincaré algebra of $(2+1)$ dimensions.

The Lorentz charges do not belong to the infinite towers of constants of motion mentioned earlier. Rather, they act as raising and lowering operators. One verifies, for the Q_n and I_n^+,

$$\begin{aligned} &\{B, I_n^+\} = -nI_{n-1}^+, && \{B, Q_n\} = -nQ_{n-1}, \\ &\{D, I_n^+\} = (n-1)I_n^+, && \{D, Q_n\} = (n - \tfrac{1}{2})Q_n, \\ &\{G, I_n^+\} = \left(\frac{n}{2} - 1\right) I_{n+1}^+, && \{G, Q_n\} = \left(\frac{n}{2} - \tfrac{1}{2}\right) Q_{n+1}. \end{aligned} \tag{6.101}$$

[Note that the $\{B, I_2^+\}$ bracket coincides with $\{B, 2H\}$, which should equal $-2P$ according to the Galileo algebra. But the above result, namely, $-2I_1^+$, gives $-2(P - \sqrt{2\lambda}\Omega)$. This central addition arises from a term of the form

$$\int dx\, dy\, \sqrt{2\lambda}\, x \frac{\partial}{\partial x} \delta(x-y)$$

whose value is ambiguous, depending on the order of integration.] The brackets with the I_n^- do not close, but the I_n^- can be modified by the addition of another tower of constant quantities, namely, those of (6.85),

$$\tilde{I}_n^- = I_n^- - \sqrt{2\lambda}\, n(n-1) \int dx\, R_-^{n-2} \psi\, \partial_x \psi. \tag{6.102}$$

The modified constants obey the same algebra as I_n^+:

$$\begin{gathered} \{B, \tilde{I}_n^-\} = -n\tilde{I}_{n-1}^-, \qquad \{D, \tilde{I}_n^-\} = (n-1)\tilde{I}_n^-, \\ \{G, \tilde{I}_n^-\} = \left(\frac{n}{2} - 1\right) \tilde{I}_{n+1}^-. \end{gathered} \tag{6.103}$$

Evidently, I_n^+, $\tilde{I}_n^-$, and Q_n provide irreducible, infinite-dimensional representations for SO(2, 1), with the Casimir, in adjoint action, taking the form $l(l+1)$, and $l=1$ for I_n^+, $\tilde{I}_n^-$, and $l=\frac{1}{2}$ for Q_n.

Finally, we inquire about the algebra of the towers of extended charges I_n^+, $\tilde{I}_n^-$, and Q_n. While some (bosonic) brackets vanish, others provide new constants of motion like those in (6.84)–(6.86) and their generalizations with more derivatives. Thus it appears that one is dealing with an open (super)algebra.

7

Toward a Non-Abelian Fluid Mechanics

Fluid mechanics and fluid magnetohydrodynamics may very well describe the long-wavelength degrees of freedom in a quark–gluon plasma. Moreover, it is plausible that the group (color) degrees of freedom remain distinct in that regime, so that one should incorporate them in the fluid approximation. In this way, one is led to think about constructing non-Abelian fluid mechanics and (color) magnetohydrodynamics. In this section we describe an approach to this task [39]. In the course of development, we encounter and solve an interesting mathematical problem: how to parametrize a non-Abelian vector potential so that the non-Abelian Chern–Simons density becomes a total derivative, and the volume-integrated Chern–Simons term is given by a surface integral. Obviously, this is the non-Abelian generalization of the similar Abelian problem, which is solved by presenting the Abelian vector potential in Clebsch form. So we shall determine the non-Abelian version of the Clebsch parametrization.

7.1 Proposal for Non-Abelian Fluid Mechanics

We review our Lagrange density for relativistic Abelian fluid mechanics (3.61):

$$\mathcal{L} = -j^\mu a_\mu - f(\sqrt{j^\mu j_\mu}). \tag{7.1}$$

The equation of state is encoded in the function f. For free fluid motion $f(\sqrt{j^\mu j_\mu}) = c\sqrt{j^\mu j_\mu}$. Here j^μ is the matter current and a_μ is an auxilliary 4-vector, which is presented in the form

$$a_\mu = \partial_\mu \theta + \alpha\, \partial_\mu \beta. \tag{7.2}$$

The time component a_0, involving time derivatives, determines the canonical 1-form; the spatial components $\boldsymbol{a}$ are in the Clebsch parametrization, as is needed for overcoming the obstacle created by a Casimir invariant of the fluid in the algebra (2.14), (2.16). Another way of characterizing the parametrization of the vector $\boldsymbol{a}$ is that it casts the Chern–Simons density of $\boldsymbol{a}$, namely, $\boldsymbol{a} \cdot \boldsymbol{\nabla} \times \boldsymbol{a}$, into total derivative form: $\boldsymbol{\nabla}\theta \cdot (\boldsymbol{\nabla}\alpha \times \boldsymbol{\nabla}\beta) = \boldsymbol{\nabla} \cdot (\theta \boldsymbol{\nabla} \times \boldsymbol{a})$.

For a non-Abelian generalization, it is plausible to suppose that the current 4-vector acquires an internal symmetry index: J^μ_a; correspondingly, the auxiliary 4-vector must also acquire an internal symmetry index: A^a_μ. It remains to give a rule for parametrizing A^a_μ, which generalizes the Abelian rule (7.2).

Our proposal—and it is a speculative one, since at this stage we have no derivation from microscopic considerations—is that A^a_μ should be written in a form so that its non-Abelian Chern–Simons density, $\mathrm{CS}(A) = A^a\, dA^a + \frac{1}{3} f^{abc} A^a A^b A^c$, is a total derivative (f^{abc} are the structure constants of the group). This leads us to the purely mathematical problem of constructing a parametrization for a non-Abelian vector that ensures this property.

7.2 Non-Abelian Clebsch Parametrization (or, Casting the Non-Abelian Chern–Simons Density into Total Derivative Form)

We enquire whether it is possible to parametrize the non-Abelian 1-form, A^a, such that the Chern–Simons 3-form is a total derivative (is exact):

$$\mathrm{CS}(A) = A^a\, dA^a + \tfrac{1}{3} f^{abc} A^a A^b A^c = d\Omega\,. \tag{7.3}$$

That this should be possible follows from the observation that the left side of (7.3) is a 3-form on 3-space; hence it is closed, because a 4-form does not exist in 3-space. [Of course, on a four-dimensional space the exterior derivative of (7.3) is proportional to the non-Abelian anomaly (Chern–Pontryagin density) [13].]

But a closed form is also exact, at least locally; this justifies the right side of (7.3).

How this works in the Abelian case has already been explored in Sections 2.4 and 2.5: the Clebsch parametrization (2.58), (2.60) for A leads to the desired result. But the generalization of (2.58), (2.60) for a non-Abelian 1-form is not evident. However, at the end of Section 2.5, an alternative approach is presented, wherein the Abelian 1-form is projected from a non-Abelian pure gauge 1-form. This construction can be generalized to the non-Abelian case and yields the sought-for parametrization.

The mathematical problem can therefore be formulated in the following way:

> *For a given group H, how can one construct a potential $A^a_\mu = (A^a_0, A^a_i)$ such that the non-Abelian Chern–Simons integrand* CS(A) *is a total derivative?*

Here we shall only sketch the solution to the problem, referring those interested to [39] for a detailed discussion.

In the solution that we present, the "total derivative" form for the Chern–Simons density of A^a is achieved in two steps. The parametrization, which we find, leads directly to an Abelian form of the Chern–Simons density

$$A^a\, dA^a + \tfrac{1}{3} f^{abc} A^a A^b A^c = \gamma\, d\gamma, \tag{7.4}$$

for some γ. Then Darboux's theorem [10] (or usual fluid dynamical theory [11]) ensures that γ can be presented in Clebsch form, so that $\gamma\, d\gamma$ is explicitly a total derivative.

We begin with a pure gauge $g^{-1}\, dg$ in some non-Abelian group G (called the *Ur*-group) whose Chern–Simons integral coincides with the winding number of g:

$$W(g) = \frac{1}{16\pi^2}\int d^3r\, \mathrm{CS}(g^{-1}\, dg) = \frac{1}{24\pi^2}\int \mathrm{tr}(g^{-1}\, dg)^3. \tag{7.5}$$

We consider a normal subgroup $H \subset G$, with generators I^a, and construct a non-Abelian gauge potential for H by projection

$$A^a \propto \mathrm{tr}(I^a g^{-1}\, dg). \tag{7.6}$$

Within H this is not a pure gauge. We determine the group structure that ensures the Chern–Simons 3-form of A^a to be proportional to $\operatorname{tr}(g^{-1}\,dg)^3$. Consequently, the constructed non-Abelian gauge fields, belonging to the group H, carry a quantized Chern–Simons number. Moreover, we describe the properties of the *Ur*-group G that guarantee that the projected potential A^a enjoys sufficient generality to represent an arbitrary potential in H.

Since $\operatorname{tr}(g^{-1}\,dg)^3$ is a total derivative for an arbitrary group (although this fact cannot in general be expressed in finite terms [40]) our construction ensures that the form of A^a, which is achieved through the projection (7.6), produces a "total derivative" expression (in the limited sense indicated above) for its Chern–Simons density.

Conditions on the *Ur*-group G, which we take to be compact and semisimple, are the following. First of all G has to be so chosen that it has a sufficient number of parameters to make $\operatorname{tr}(I^a g^{-1}\,dg)$ a generic potential for H. Since we are in three dimensions, an H-potential A_i^a has $3 \times \dim H$ independent functions; so a minimal requirement will be

$$\dim G \geq 3 \dim H. \tag{7.7}$$

Second, we require that the H-Chern–Simons form for A^a should coincide with that of $g^{-1}\,dg$. As we shall show in a moment, this is achieved if G/H is a symmetric space. In this case, if we split the Lie algebra of G into the H-subalgebra spanned by I^a ($a = 1, \ldots, \dim H$), and the orthogonal complement spanned by S^A ($A = 1, \ldots, (\dim G - \dim H)$), the commutation rules are of the form

$$[I^a, I^b] = f^{abc} I^c, \tag{7.8a}$$
$$[I^a, S^A] = h^{aAB} S^B, \tag{7.8b}$$
$$[S^A, S^B] = N h^{aAB} I^a, \tag{7.8c}$$

$(h^a)^{AB}$ form a (possibly reducible) representation of the H-generators I^a. The constant N depends on normalizations. More explicitly, if the structure constants for the *Ur*-group G are named $\bar{f}^{abc}$ $(a, b, c = 1, \ldots, \dim G)$, then the conditions (7.8a)–(7.8c) require that $\bar{f}^{abc}$ vanishes whenever an odd number of indices belongs to the orthogonal complement labeled by A, $B, \ldots$. Moreover, f^{abc} are taken to be the conventional structure constants for H and this may render them proportional to (rather than equal to) $\bar{f}^{abc}$.

We define the traces of the generators by

$$\begin{gathered} \operatorname{tr}(I^a I^b) = -N_1 \delta^{ab}, \qquad \operatorname{tr}(S^A S^B) = -N_2 \delta^{AB}, \\ \operatorname{tr}(I^a S^A) = 0. \end{gathered} \tag{7.9}$$

We can evaluate the quantity $\operatorname{tr}[S^A, S^B] I^a = \operatorname{tr} S^A [S^B, I^a]$ using the commutation rules. This immediately gives the relation $N_1 N = N_2$.

Expanding $g^{-1}\, dg$ in terms of generators, we write

$$g^{-1}\, dg = (I^a A^a + S^A \alpha^A), \tag{7.10}$$

which defines the H-potential A^a. Equivalently,

$$A^a = -\frac{1}{N_1} \operatorname{tr}(I^a g^{-1}\, dg). \tag{7.11}$$

From $d(g^{-1}\, dg) = -g^{-1}\, dg\, g^{-1}\, dg$, we get the Maurer–Cartan relations

$$\begin{aligned} F^\alpha \equiv dA^a + \tfrac{1}{2} f^{abc} A^b A^c &= -\frac{N}{2} h^{aAB} \alpha^A \alpha^B, \\ d\alpha^A + h^{\alpha BA} A^a \alpha^B &= 0. \end{aligned} \tag{7.12}$$

Using these results, the following chain of equations shows that the Chern–Simons 3-form for the H-gauge group is proportional to $\operatorname{tr}(g^{-1}\, dg)^3$:

$$\begin{aligned} &\frac{1}{16\pi^2}(A^a\, dA^a + \tfrac{1}{3} f^{abc} A^a A^b A^c) \\ &\qquad = \frac{1}{48\pi^2}(A^a\, dA^a + 2 A^a F^a) \\ &\qquad = \frac{1}{48\pi^2}(A^a\, dA^a - N h^{aAB} A^a \alpha^A \alpha^B) \end{aligned}$$

$$
\begin{aligned}
&= \frac{1}{48\pi^2}(A^a\, dA^a + N\, d\alpha^A\, \alpha^A) \\
&= -\frac{1}{48\pi^2}\left[\frac{1}{N_1}\operatorname{tr}(A\, dA) + \frac{N}{N_2}\operatorname{tr}(d\alpha\, \alpha)\right] \\
&= -\frac{1}{48\pi^2 N_1}\operatorname{tr}(A\, dA + \alpha\, d\alpha) \\
&= -\frac{1}{48\pi^2 N_1}\operatorname{tr} g^{-1}\, dg\, d(g^{-1}\, dg) \\
&= \frac{1}{48\pi^2 N_1}\operatorname{tr}(g^{-1}\, dg)^3. \qquad (7.13)
\end{aligned}
$$

In the above sequence of manipulations, we have used the Maurer–Cartan relations (7.12), which rely on the symmetric space structure of (7.8a)–(7.8c), and the trace relations (7.9), along with $N_1 N = N_2$.

We thus see that $\int \mathrm{CS}(A)$ is indeed the winding number of the configuration $g \in G$. Since $\operatorname{tr}(g^{-1}\, dg)^3$ is a total derivative locally on G, the potential (7.11), with the symmetric space structure of (7.8a)–(7.8c), does indeed fulfill the requirement of making $\mathrm{CS}(A)$ a total derivative. It is therefore appropriate to call our construction (7.11) a "non-Abelian Clebsch parametrization."

In explicit realizations, given a gauge group of interest H, we need to choose a group G such that the conditions (7.7), (7.8a)–(7.8c) hold. In general, this is not possible. However, one can proceed recursively. Let us suppose that the desired result has been established for a group, which we call H_2. Then we form $H \subset G$ obeying (7.8a)–(7.8c) as $H = H_1 \times H_2$, where H_1 is the gauge group of interest, satisfying $\dim G \geq 3 \dim H_1$. For this choice of H, the result (7.13) becomes

$$
\mathrm{CS}(H_1) + \mathrm{CS}(H_2) = \frac{1}{48\pi^2 N_1}\operatorname{tr}(g^{-1}\, dg)^3. \qquad (7.14)
$$

But since $\mathrm{CS}(H_2)$ is already known to be a total derivative, (7.14) shows the desired result: $\mathrm{CS}(H_1)$ is a total derivative.

To see explicitly how this works we work out the representation for an $\mathrm{SU}(2) \approx \mathrm{O}(3)$ potential A_i^a, which possesses nine independent functions.

We take $G = \mathrm{O}(5)$, $H = \mathrm{O}(3) \times \mathrm{O}(2)$. We consider the four-dimensional spinorial representation of O(5). With the generators normalized by $\operatorname{tr}(t^a t^b) = -\delta^{ab}$, the Lie algebra generators of O(5) are given by

$$\begin{aligned} I^a &= \frac{1}{2i}\begin{pmatrix} \sigma^a & 0 \\ 0 & \sigma^a \end{pmatrix}, & I^0 &= \frac{1}{2i}\begin{pmatrix} -1 & 0 \\ 0 & 1 \end{pmatrix}, \\ S^A &= \frac{1}{i\sqrt{2}}\begin{pmatrix} 0 & 0 \\ \sigma^A & 0 \end{pmatrix}, & \widetilde{S}^A &= \frac{1}{i\sqrt{2}}\begin{pmatrix} 0 & \sigma^A \\ 0 & 0 \end{pmatrix}, \end{aligned} \tag{7.15}$$

where σ's are the 2×2 Pauli matrices. I^a generate O(3), with the conventional structure constants ε^{abc}, and I^0 is the generator of O(2). S, $\widetilde{S}$ are the coset generators.

A general group element in O(5) can be written in the form $g = Mhk$ where $h \in \mathrm{O}(3)$, $k \in \mathrm{O}(2)$, and

$$\begin{aligned} M = &\frac{1}{\sqrt{1 + \overline{\boldsymbol{w}} \cdot \boldsymbol{w} - \frac{1}{4}(\boldsymbol{w} \times \overline{\boldsymbol{w}})^2}} \\ &\times \begin{pmatrix} 1 - (i/2)(\boldsymbol{w} \times \overline{\boldsymbol{w}}) \cdot \boldsymbol{\sigma} & -\boldsymbol{w} \cdot \boldsymbol{\sigma} \\ \overline{\boldsymbol{w}} \cdot \boldsymbol{\sigma} & 1 + (i/2)(\boldsymbol{w} \times \overline{\boldsymbol{w}}) \cdot \boldsymbol{\sigma} \end{pmatrix}, \end{aligned} \tag{7.16}$$

w^a is a complex three-dimensional vector, with the bar denoting complex conjugation. $\boldsymbol{w} \cdot \overline{\boldsymbol{w}} = w^a \bar{w}^a$ and $(\boldsymbol{w} \times \overline{\boldsymbol{w}})^a = \varepsilon^{abc} w^b \bar{w}^c$. The general O(5) group element contains ten independent real functions. These are collected as six from M (in the three complex functions w^a), three in h, and one in k.

The O(3) gauge potential given by $-\operatorname{tr}(I^a g^{-1}\, dg)$ reads

$$\begin{aligned} A^a &= R^{ab}(h) a^b + (h^{-1}\, dh)^a, \\ a^a &= \frac{1}{1 + \boldsymbol{w} \cdot \overline{\boldsymbol{w}} - \frac{1}{4}(\boldsymbol{w} \times \overline{\boldsymbol{w}})^2} \\ &\quad \times \left\{ \frac{w^a\, d\overline{\boldsymbol{w}} \cdot (\boldsymbol{w} \times \overline{\boldsymbol{w}}) + \bar{w}^a\, d\boldsymbol{w} \cdot (\overline{\boldsymbol{w}} \times \boldsymbol{w})}{2} \right. \\ &\qquad \left. + \varepsilon^{abc}(dw^b\, \bar{w}^c - w^b\, d\bar{w}^c) \right\}, \end{aligned} \tag{7.17}$$

where $R^{ab}(h)$ is defined by $hI^a h^{-1} = R^{ab} h^b$ and k does not contribute. A^a is the h-gauge transform of a^a, which depends on six real parameters (w^a). The three gauge parameters of $h \in \mathrm{O}(3)$, along with the six, give the nine functions needed to parametrize a general O(3)- [or SU(2)-] potential in three dimensions. The Chern–Simons form is

$$\begin{aligned}\mathrm{CS}(A) &= \frac{1}{16\pi^2}\left(A^a\, dA^a + \tfrac{1}{3}\varepsilon^{abc} A^a A^b A^c\right),\\ &= \frac{1}{16\pi^2}\left(a^a\, da^a + \tfrac{1}{3}\varepsilon^{abc} a^a a^b a^c\right)\\ &\quad - d\left[\frac{1}{16\pi^2}(dh\, h^{-1})^a a^a)\right] + \frac{1}{24\pi^2}\,\mathrm{tr}(h^{-1}\, dh)^3. \qquad (7.18)\end{aligned}$$

The second equality reflects the usual response of the Chern–Simons density to gauge transformations. Using the explicit form of a^a as given in (7.17), we can further reduce this. Indeed we find that

$$\begin{aligned} a^a\, da^a + \tfrac{1}{3}\varepsilon^{abc} a^a a^b a^c &= (-2)\frac{(\overline{\boldsymbol{w}} \times d\overline{\boldsymbol{w}}) \cdot \boldsymbol{\rho} + (\boldsymbol{w} \times d\boldsymbol{w}) \cdot \overline{\boldsymbol{\rho}}}{[1 + \boldsymbol{w}\cdot\overline{\boldsymbol{w}} - \frac{1}{4}(\boldsymbol{w}\times\overline{\boldsymbol{w}})^2]^2}, \qquad (7.19)\\ \rho_k &= \tfrac{1}{2}\varepsilon_{ijk}\, d\bar{w}^i\, d\bar{w}^j\, . \end{aligned}$$

Defining an Abelian potential

$$a = \frac{\boldsymbol{w} \cdot d\overline{\boldsymbol{w}} - \overline{\boldsymbol{w}} \cdot d\boldsymbol{w}}{1 + \boldsymbol{w} \cdot \overline{\boldsymbol{w}} - \frac{1}{4}(\boldsymbol{w} \times \overline{\boldsymbol{w}})^2}, \qquad (7.20)$$

we can easily check that $a\, da$ reproduces (7.19). In other words

$$\begin{aligned}\mathrm{CS}(A) &= \frac{1}{16\pi^2} a\, da + d\left[\frac{(dh\, h^{-1})^a a^a)}{16\pi^2}\right]\\ &\quad + \frac{1}{48\pi^2}\,\mathrm{tr}(h^{-1}\, dh)^3. \qquad (7.21)\end{aligned}$$

If desired, the Abelian potential a can now be written in the Clebsch form making $a\, da$ into a total derivative, while the remaining two terms already are total derivatives, though in a "hidden" form for the last expression. This completes our construction.

7.3 Proposal for Non-Abelian Magnetohydrodynamics

We return to our construction of a Lagrange density for non-Abelian kinetic theory. As explained in Section 7.1, a plausible non-Abelian generalization for (7.1) is

$$\mathcal{L} = -J^{\mu a}\frac{1}{N_1}\operatorname{tr} I^a g^{-1}\,\partial_\mu g - c\sqrt{J^{\mu a}J^a{}_\mu}, \tag{7.22}$$

where for simplicity we have taken the "free" form for f. When the desired group is SU(2), g is an O(5) group element, as detailed in Section 7.2.

Magnetohydrodynamics is achieved by introducing a further interaction with a dynamical gauge potential $\mathcal{A}^a_\mu$. This is accomplished by promoting the derivative of g to a gauge-covariant derivative, gauged on the right

$$\mathcal{L}_{\text{magnetohydrodynamics}} = -\frac{1}{N_1}J^{\mu a}\operatorname{tr}(I^a g^{-1}D_\mu g) - c\sqrt{J^{\mu a}J^a_\mu} - \tfrac{1}{4}\mathcal{F}^{a\mu\nu}\mathcal{F}^a_{\mu\nu}, \tag{7.23}$$

with

$$D_\mu g = \partial_\mu g + eg\mathcal{A}_\mu. \tag{7.24}$$

where $\mathcal{A}_\mu = \mathcal{A}^a_\mu I^a$ are independent, dynamical gauge potentials (not given by g) leading to the field strengths $\mathcal{F}^a_{\mu\nu}$. The gauge transformation properties by the gauge function h are

$$\begin{gathered} g' = gh, \qquad \mathcal{A}' = h^{-1}\mathcal{A}h + \frac{1}{e}h^{-1}\,dh, \\ J'^a_\mu I^a = h^{-1}J^a_\mu I^a h. \end{gathered} \tag{7.25}$$

We expect that the Lagrangian (7.23) will describe non-Abelian magnetohydrodynamics, namely, the dynamics of a fluid with non-Abelian charge coupled to non-Abelian fields. [The Abelian version of (7.23) does indeed describe ordinary magnetohydrodynamics.] This gluon hydrodynamics can be useful for non-Abelian plasmas such as the quark–gluon plasma. Details of (7.23) and possible applications are under further study.

8

Noncommutative Fluid Mechanics

A focus of recent research is the suggestion that spatial coordinates need not commute. Models have been constructed that incorporate this, within both particle and field physics. The present attention to this idea derives from string theory but, in fact, a more immediately appreciated physical setting is available: motion in a strong, external magnetic field. In particular, a charged fluid interacting with such a field exhibits noncommutative features, which we discuss in this chapter. But first we show how the noncommutativity of particle coordinates (as opposed to fields) manifests itself in an analogous magnetic system.

Like many interesting quantal ideas, the notion that spatial coordinates may not commute can be traced to Heisenberg who, in a letter to Peierls, suggested that a coordinate uncertainty principle may ameliorate the problem of infinite self-energies. We shall describe later the physical application that Peierls made with Heisenberg's idea. Evidently, Peierls also described it to Pauli, who told it to Oppenheimer, who told it to Snyder, who then wrote the first paper on the subject [41].

Let us begin with a physical application of the idea that goes back to Peierls.

8.1 Noncommuting Particles

We are interested in a point-particle moving on a plane, with an external magnetic field $\mathbf{b}$ perpendicular to the plane. The equation for the 2-vector $\boldsymbol{r} = (x, y)$ is

$$m\dot{v}^i = \frac{e}{c}\varepsilon^{ij} v^j b + f^i(\boldsymbol{r}) \tag{8.1}$$

where $\mathbf{v}$ is the velocity $\dot{\boldsymbol{r}}$, and $\mathbf{f}$ represents other forces, which we take to be derived from a potential V: $\boldsymbol{f} = -\boldsymbol{\nabla} V$. Absent additional forces, the quantized theory gives rise to the well-known Landau levels, with separations $O(b/m)$. The limit of large b effectively projects onto the lowest Landau level, and is equivalent to small m. Setting the mass to zero in (8.1) leaves a first-order equation

$$\dot{r}^i = \frac{c}{eb} \varepsilon^{ij} f^j(\mathbf{r}). \tag{8.2}$$

This may be obtained by taking Poisson brackets of $\mathbf{r}$ with the Hamiltonian

$$H_0 = V \tag{8.3}$$

provided the fundamental brackets describe noncommuting coordinates

$$\{r^i, r^j\} = \frac{c}{eb} \varepsilon^{ij} \tag{8.4}$$

so that

$$\dot{r}^i = \{H_0, r^i\} = \{r^j, r^i\} \, \partial_j V = \frac{c}{eb} \varepsilon^{ij} f^j(\mathbf{r}). \tag{8.5}$$

The noncommutative algebra (8.4) and the associated dynamics can be derived in the following manner. The Lagrangian for the equation of motion (8.1) is

$$L = \tfrac{1}{2} m v^2 + \frac{e}{c} \boldsymbol{v} \cdot \boldsymbol{A} - V, \tag{8.6}$$

where we choose the gauge $\boldsymbol{A} = (0, bx)$. Setting m to zero leaves

$$L_0 = \frac{eb}{c} x \dot{y} - V(x, y), \tag{8.7}$$

which is of the form $p\dot{q} - h(p, q)$, and one sees that $\big((eb/c)x, y\big)$ form a canonical pair. This implies (8.4), and identifies V as the Hamiltonian [42].

Finally, we give a canonical derivation of noncommutativity in the $m \to 0$ limit, starting with the Hamiltonian

$$H = \frac{\boldsymbol{\pi}^2}{2m} + V, \tag{8.8}$$

where H gives (8.1) upon bracketing with $\boldsymbol{r}$ and $\boldsymbol{\pi}$, provided the following brackets hold:

$$\{r^i, r^j\} = 0, \tag{8.9}$$

$$\{r^i, \pi^j\} = \delta^{ij}, \tag{8.10}$$

$$\{\pi^i, \pi^j\} = -\frac{eb}{c}\varepsilon^{ij}. \tag{8.11}$$

Here $\boldsymbol{\pi}$ is the kinematical (noncanonical) momentum, $m\dot{\boldsymbol{r}}$, related to the canonical momentum $\boldsymbol{p}$ by $\boldsymbol{\pi} = \boldsymbol{p} - (e/c)\boldsymbol{A}$.

We wish to set m to zero in (8.8). This can only be done provided $\boldsymbol{\pi}$ vanishes, and we impose $\boldsymbol{\pi} = 0$ as a constraint. But according to (8.11), the bracket of the constraints is nonzero, and the constraints are recognized to be "second-class" in Dirac's terminology. To proceed with the canonical formalism, we must introduce Dirac brackets. We omit the details of that technology, but merely record the resulting Dirac bracket [42]:

$$\{r^i, r^j\}_D = \frac{c}{eb}\varepsilon^{ij}. \tag{8.12}$$

Evidently, in this approach, noncommuting coordinates arise as Dirac brackets in a system constrained to lie in the lowest Landau level. Notice that the coordinate noncommutativity is already established at the classical level in that the Poisson bracket of coordinates is nonvanishing. Later we shall discuss the quantum version.

Peierls observed that when an impurity in the electron system is described by V, one can obtain the first-order energy shift of the lowest Landau level by taking the coordinates of (x, y) on which V depends to be noncommuting [43].

A further interesting subject, which is not discussed here, concerns the behavior of the wave function in the phase-space reductive, $m \to 0$, a limit that projects onto the lowest Landau level. Before the reduction, the wave function is a normalized expression depending on the two coordinates (x, y). After the reduction, the wave function can depend only on one coordinate, because the other is a conjugate variable. How all this comes about is explained in the literature [42].

8.2 Noncommuting Fluids

The above demonstrates that spatial coordinates of particles in an intense magnetic field do not (Poisson) commute. But we are interested in fields. To find an example of noncommuting fields, we turn to the equations of a charged fluid, moving on a plane in an external magnetic field perpendicular to the plane. The fluid is described by a density ρ and velocity $\boldsymbol{v}$, both defined on the two-dimensional plane. A mass parameter m is introduced for dimensional reasons, so that the mass density is $m\rho$. The fields ρ and $\boldsymbol{v}$ are functions of t and $\boldsymbol{r}$ and give an Eulerian description of the fluid. The equations that are satisfied are the continuity equation

$$\dot{\rho} + \boldsymbol{\nabla} \cdot (\rho \boldsymbol{v}) = 0, \tag{8.13}$$

which expresses matter conservation, and the Euler equation

$$m\dot{v}^i + m\boldsymbol{v} \cdot \boldsymbol{\nabla} v^i = \frac{e}{c}\varepsilon^{ij} v^j b + f^i, \tag{8.14}$$

which is the force equation. Here f^i describes additional forces, e.g., $-\rho^{-1}\boldsymbol{\nabla} P$ where P is pressure. We shall take the force to be derived from a potential of the form

$$\boldsymbol{f}(\boldsymbol{r}) = -\boldsymbol{\nabla} \frac{\delta}{\delta\rho(\boldsymbol{r})} \int d\boldsymbol{r}\, V, \tag{8.15}$$

where we allow a general dependence of V on ρ (e.g., nonlocality or dependence on derivatives of ρ) and also a translation noninvariant, explicit dependence on $\boldsymbol{r}$ [compare (2.1)–(2.3)].

Equations (8.13)–(8.15) follow by bracketing ρ and $\boldsymbol{\pi}$ with the Hamiltonian

$$H = \int d^2r \left(\rho \frac{\boldsymbol{\pi}^2}{2m} + V \right) \tag{8.16}$$

provided that fundamental brackets are taken as

$$\{\rho(\boldsymbol{r}), \rho(\boldsymbol{r}')\} = 0, \tag{8.17}$$

$$\{\pi(\boldsymbol{r}), \rho(\boldsymbol{r}')\} = \boldsymbol{\nabla}\delta(\boldsymbol{r} - \boldsymbol{r}'), \tag{8.18}$$

$$\{\pi^i(\boldsymbol{r}), \pi^j(\boldsymbol{r}')\} = -\varepsilon^{ij} \frac{1}{\rho(\boldsymbol{r})} \left(m\omega(\boldsymbol{r}) + \frac{eb}{c} \right) \delta(\boldsymbol{r} - \boldsymbol{r}'), \tag{8.19}$$

where $\varepsilon^{ij}\omega(\boldsymbol{r})$ is the vorticity $\partial_i v^j - \partial_j v^i$, and $\boldsymbol{\pi} = m\boldsymbol{v}$ [compare (2.11)–(2.14)].

We now consider a strong magnetic field and take the limit $m \to 0$, which is equivalent to large b. Equations (8.14) and (8.15) reduce to

$$v^i = -\frac{c}{eb}\varepsilon^{ij}\frac{\partial}{\partial r^j}\frac{\delta}{\delta\rho(\boldsymbol{r})}\int d^2r\, V. \tag{8.20}$$

Combining this with the continuity equation (8.13) gives the equation for the density "in the lowest Landau level":

$$\dot{\rho}(\boldsymbol{r}) = \frac{c}{eb}\frac{\partial}{\partial r^i}\rho(\boldsymbol{r})\varepsilon^{ij}\frac{\partial}{\partial r^j}\frac{\delta}{\delta\rho(\boldsymbol{r})}\int d^2r\, V. \tag{8.21}$$

(For the right-hand side not to vanish, V must not be solely a function of ρ.)

The equation of motion (8.21) can be obtained by bracketing with the Hamiltonian

$$H_0 = \int d^2r\, V \tag{8.22}$$

provided the charge density bracket is nonvanishing, showing noncommutativity of the ρ's [44]:

$$\{\rho(\boldsymbol{r}), \rho(\boldsymbol{r}')\} = -\frac{c}{eb}\varepsilon^{ij}\,\partial_i\rho(\boldsymbol{r})\,\partial_j\delta(\boldsymbol{r}-\boldsymbol{r}'). \tag{8.23}$$

H_0 and this bracket may be obtained from (8.16) and (8.17)–(8.19) with the same Dirac procedure presented for the particle case: We wish to set m to zero in (8.16); this is possible only if $\boldsymbol{\pi}$ is constrained to vanish. But the bracket of the $\boldsymbol{\pi}$'s (8.19) is nonvanishing, even at $m = 0$, because $b \neq 0$. Thus at $m = 0$ we are dealing with a second-class constraint which leads to a nonvanishing Dirac bracket of densities as in (8.23):

$$\{\rho(\boldsymbol{r}), \rho(\boldsymbol{r}')\}_D = -\frac{c}{eb}\varepsilon^{ij}\,\partial_i\rho(\boldsymbol{r})\,\partial_j\delta(\boldsymbol{r}-\boldsymbol{r}'). \tag{8.24}$$

The ρ bracket (8.23), (8.24) enjoys a more appealing expression in momentum space. Upon defining

$$\tilde{\rho}(\boldsymbol{p}) = \int d^2r\, e^{i\boldsymbol{p}\cdot\boldsymbol{r}}\rho(\boldsymbol{r}) \tag{8.25}$$

we find

$$\{\tilde{\rho}(\boldsymbol{p}), \tilde{\rho}(\boldsymbol{q})\} = -\frac{c}{eb}\varepsilon^{ij}p^i q^j \tilde{\rho}(\boldsymbol{p}+\boldsymbol{q}). \tag{8.26}$$

The form of the charge density bracket (8.23), (8.24), (8.26) can be understood by reference to the particle substructure for the fluid. Take

$$\rho(\boldsymbol{r}) = \sum_n \delta(\boldsymbol{r} - \boldsymbol{r}_n), \tag{8.27}$$

where n labels the individual particles. When the coordinates of each particle satisfy the nonvanishing bracket (8.4), (8.12), the $\{\rho(\boldsymbol{r}), \rho(\boldsymbol{r}')\}$ bracket takes the form (8.23), (8.24), (8.26).

8.3 Quantization

Quantization before the reduction to the lowest Landau level is straightforward. For the particle case (8.9)–(8.11) and for the fluid case (8.17)–(8.19) we replace brackets with $i/\hbar$ times commutators. After reduction to the lowest Landau level we do the same for the particle case thereby arriving at the "Peierls substitution," which (as mentioned previously) states that the effect of an impurity [V in (8.6)] on the lowest Landau energy level can be evaluated to lowest order by viewing the (x, y) arguments of V as noncommuting variables [43].

For the fluid, quantization presents a choice. On one hand, we can simply promote the bracket (8.23), (8.24), (8.26) to a commutator by multiplying by $i/\hbar$:

$$[\rho(\boldsymbol{r}), \rho(\boldsymbol{r}')] = i\hbar\frac{c}{eb}\varepsilon^{ij}\partial_i\rho(\boldsymbol{r}')\partial_j\delta(\boldsymbol{r}-\boldsymbol{r}'), \tag{8.28}$$

$$[\tilde{\rho}(\boldsymbol{p}), \tilde{\rho}(\boldsymbol{q})] = i\hbar\frac{c}{eb}\varepsilon^{ij}p^i q^j \tilde{\rho}(\boldsymbol{p}+\boldsymbol{q}). \tag{8.29}$$

Alternatively, we can adopt expression (8.27), for the operator $\rho(\boldsymbol{r})$, where the $\boldsymbol{r}_n$ now satisfy the noncommutative algebra

$$[r_n^i, r_{n'}^j] = -i\hbar\frac{c}{eb}\varepsilon^{ij}\delta_{nn'} \tag{8.30}$$

and calculate the ρ commutator as a derived quantity.

However, once $\boldsymbol{r}_n$ is a noncommuting operator, functions of $\boldsymbol{r}_n$, even δ-functions, have to be ordered. We choose the Weyl ordering, which is equivalent to defining the Fourier transform as

$$\tilde{\rho}(\boldsymbol{p}) = \sum_n e^{i\boldsymbol{p}\cdot\boldsymbol{r}_n}. \tag{8.31}$$

With the help of (8.30) and the Baker–Hausdorff lemma, we arrive at the "trigonometric algebra"

$$[\tilde{\rho}(\boldsymbol{p}), \tilde{\rho}(\boldsymbol{q})] = 2i \sin\left(\frac{\hbar c}{2eb}\varepsilon^{ij} p^i q^j\right)\tilde{\rho}(\boldsymbol{p}+\boldsymbol{q}). \tag{8.32}$$

This reduces to (8.29) for small $\hbar$.

This form for the commutator, (8.32), is connected to a Moyal star product in the following fashion. For an arbitrary c-number function $f(\boldsymbol{r})$ define

$$\langle f \rangle = \int d^2r\, \rho(\boldsymbol{r}) f(\boldsymbol{r}) = \frac{1}{(2\pi)^2} \int d^2p\, \tilde{\rho}(\boldsymbol{p}) \tilde{f}(-\boldsymbol{p}). \tag{8.33}$$

Multiplying (8.32) by $\tilde{f}(-\boldsymbol{p})\tilde{g}(-\boldsymbol{q})$ and integrating gives

$$[\langle f \rangle, \langle g \rangle] = \langle h \rangle \tag{8.34}$$

with

$$h(\boldsymbol{r}) = (f \star g)(\boldsymbol{r}) - (g \star f)(\boldsymbol{r}), \tag{8.35}$$

where the "$\star$" product is defined as

$$(f \star g)(\boldsymbol{r}) = e^{i\hbar c \varepsilon^{ij} \partial_i \partial'_j/(2eb)} f(\boldsymbol{r}) g(\boldsymbol{r}')|_{\boldsymbol{r}'=\boldsymbol{r}}. \tag{8.36}$$

Note however that only the commutator is mapped into the star commutator. The product $\langle f \rangle \langle g \rangle$ is not equal to $\langle f \star g \rangle$.

The lack of consilience between (8.29) and (8.32) is an instance of the Groenwald–VanHove theorem which establishes the impossibility of taking over into quantum mechanics all the classical brackets [45]. Equations (8.29)–(8.36) explicitly exhibit the physical occurrence of the star product for fields in a strong magnetic background. This product is then used in the construction of noncommuting field theories.

8.4 Various Algebras

Let us summarize the various (nontrivial) algebras that we have encountered. The velocity algebra of fluids (8.19), with $b = 0$ and $m = 1$ [see (2.14)], as noted before, is due to Landau [3]. Its relative, the momentum density algebra (2.16), also describes diffeomorphisms of space (in any number of dimensions) in the following fashion. If an infinitesimal coordinate transformation is given by

$$\delta r^i = -f^i(\boldsymbol{r}) \tag{8.37}$$

we define the average $\langle f \rangle$ of f^i by integrating with $\mathcal{P}^i$:

$$\langle f \rangle \equiv \int d\boldsymbol{r}\, f^i(\boldsymbol{r}) P^i(\boldsymbol{r}), \tag{8.38}$$

then (2.16) has the consequence that, for two such functions f and g, we have

$$\{\langle f \rangle, \langle g \rangle\} = -\langle h \rangle, \tag{8.39}$$

where h is the Lie bracket of f and g:

$$h^i = g^j\, \partial_j f^i - f^j\, \partial_j g^i. \tag{8.40}$$

By scaling ρ, the noncommutative density algebra (8.23), (8.24) may be presented as

$$\{\rho(\boldsymbol{r}), \rho(\boldsymbol{r}')\} = \varepsilon^{ij}\, \partial_i \rho(\boldsymbol{r})\, \partial_j \delta(\boldsymbol{r} - \boldsymbol{r}'). \tag{8.41}$$

This intrinsically two-dimensional structure is the area-preserving algebra, studied by Arnold [46]. Area-preserving coordinate transformations (volume preserving in arbitrary dimensionality) possess a unit Jacobian. For the infinitesimal form of the transformation (8.37) this means that f^i is transverse: $\partial_i f^i = 0$. Therefore, in two dimensions, an area-preserving transformation is generated by a scalar

$$f^i = \varepsilon^{ij}\, \partial_j f. \tag{8.42}$$

When an average $\langle f \rangle$ is defined by

$$\langle f \rangle = \int d^2 r\, f(\boldsymbol{r}) \rho(\boldsymbol{r}) \tag{8.43}$$

equation (8.41) again implies (8.39), but now we have

$$h = \varepsilon^{ij}\, \partial_i f\, \partial_j g \tag{8.44}$$

which also follows from (8.40) when all three functions take the form (8.42).

Finally, the algebra (8.32)

$$\{\tilde{\rho}(\boldsymbol{p}), \tilde{\rho}(\boldsymbol{q})\} = -\frac{2}{\hbar} \sin\left(\frac{\hbar c}{2eb} \varepsilon^{ij} p^i q^j\right) \tilde{\rho}(\boldsymbol{p}+\boldsymbol{q}),$$

which also leads to the Moyal-star product (8.36) for averages (8.43), is called a trigonometric algebra, which was introduced by D. Fairlie, P. Fletcher, and C. Zachos [47].

Solutions to Problems

Problem 2.1. The imaginary part of the Schrödinger equation gives the continuity equation in the form $\dot{\rho} + \boldsymbol{\nabla} \cdot (\rho \boldsymbol{\nabla} \theta) = 0$. This identifies the velocity $\boldsymbol{v}$ as $\boldsymbol{\nabla}\theta$, that is, $\boldsymbol{v}$ is irrotational and there is no vorticity. The real part becomes the Bernoulli equation $\dot{\theta} + \frac{1}{2}(\boldsymbol{\nabla}\theta)^2 = \frac{1}{2}\hbar^2 \rho^{-1/2} \nabla^2 \rho^{1/2}$, whose gradient gives the Euler equation and identifies the force $\boldsymbol{f}$ as $\boldsymbol{\nabla}(\frac{1}{2}\hbar^2 \rho^{-1/2} \nabla^2 \rho^{1/2})$.

Problem 2.2. $\boldsymbol{j} = \rho \boldsymbol{v}$ with $\boldsymbol{v} = \boldsymbol{\nabla}\theta$.

Problem 2.3. $\mathcal{L}_{\text{Schrödinger}} = \theta\dot{\rho} - \frac{1}{2}\rho(\nabla\theta)^2 - \hbar^2/8 \boldsymbol{\nabla}\rho \cdot \boldsymbol{\nabla}\rho/\rho$, where the time derivative of $i\hbar\rho/2 - \rho\theta$ has been dropped.

The results in the solutions to Problems 2.1 and 2.3 are called the Mädelung formulation of quantum mechanics [48].

Problem 2.4. $\text{CS}(A) = \varepsilon^{ijk}\, \partial_i \Phi\, \partial_j \cos\Theta \partial_k h(r)$

(a) Extracting the first derivative leaves

$$\text{CS}(A) = \partial_i V_a^i, \qquad V_a^i = \varepsilon^{ijk} \Phi \partial_j \cos\Theta\, \partial_k h(r).$$

(This is true because $\varepsilon^{ijk}\, \partial_i \partial_j \cos\Theta = 0 = \varepsilon^{ijk}\, \partial_i \partial_k h(r)$, since $\cos\Theta$ and $h(r)$ are nonsingular.) Note that

$$V_a^i = \varepsilon^{i\Theta r} \Phi(-\sin\Theta/r) h'(r) = \delta^{i\Phi} \Phi(1/r \sin\Theta) h(r).$$

Since $V_a^r = 0$, the surface integral does not contribute. However, since Φ is multivalued, there is a contribution from the Φ integral:

$$\begin{aligned}
&\int d^3 r\, \text{CS}(A) \\
&= \int_0^R r^2\, dr \int_0^\pi \sin\Theta\, d\Theta \int_0^{2\pi} d\Phi \left(\frac{1}{r \sin\Theta} \frac{\partial}{\partial \Phi} \Phi \right) \left(\frac{1}{r} \sin\Theta \right) h'(r) \\
&= 4\pi [h(R) - h(0)].
\end{aligned}$$

(b) Extracting the second derivative leaves

$$\mathrm{CS}(A) = \partial_j V_b^j - \varepsilon^{ijk}(\partial_j \partial_i \Phi) \cos\Theta\, \partial_k h(r).$$

The last term is present, owing to the singularity of Φ at the origin, which gives rise to $\varepsilon^{kij}\partial_i\partial_j\Phi = \delta^{k3}2\pi\delta(x)\delta(y)$. (See [49].) Also we have

$$\begin{aligned} V_b^i = \varepsilon^{ijk}\, \partial_i \Phi \cos\Theta\, \partial_k h(r) &= \varepsilon^{\Phi jr}\left(\frac{1}{r\sin\theta}\right) \cos\Theta h'(r) \\ &= -\frac{1}{r}\delta^{j\theta} \cot\Theta h'(r). \end{aligned}$$

Again there is no r-component to contribute to the surface integral, but the second, singular term leaves

$$\begin{aligned} \int d^3r\, \mathrm{CS}(A) &= \int d^3r (2\pi)\delta(x)\delta(y) \cos\Theta \frac{\partial}{\partial z} h(r) \\ &= 2\pi \int_{-R}^{R} dz\, \frac{z}{|z|}\frac{\partial}{\partial z} h(|z|) \\ &= 4\pi \int_0^R dz\, \frac{\partial}{\partial z} h(z) = 4\pi\big[h(R) - h(0)\big]. \end{aligned}$$

(c) Extracting the last derivative leaves

$$\begin{aligned} \mathrm{CS}(A) &= \partial_k V_c^k - \varepsilon^{ijk}(\partial_k\partial_i\Phi)\partial_j \cos\Theta h(r), \\ V_c^k &= \varepsilon^{ijk}\, \partial_i\Phi\partial_j \cos\Theta h(r) \\ &= \varepsilon^{\Phi\theta k}\left(\frac{1}{r\sin\Theta}\right)\left(-\frac{1}{r}\sin\Theta\right) h(r) \\ &= \delta^{kr}\frac{h(r)}{r^2}. \end{aligned}$$

Here the surface integral contributes $4\pi h(R)$. The singular term is

$$\begin{aligned} -\delta^{j3}(2\pi)\delta(x)\delta(y)(\partial_j \cos\Theta)h(r) &= -(2\pi)\delta(x)\delta(y)\left(\frac{\partial}{\partial z}\frac{z}{|z|}\right)h(|z|) \\ &= -(4\pi)\delta^3(\boldsymbol{r})h(0). \end{aligned}$$

Hence this contribution to the spatial integral is $-4\pi h(0)$, for a total of $4\pi\big[h(R) - h(0)\big]$.

Problem 2.5. In the Clebsch parameterization, $\boldsymbol{B} = \boldsymbol{\nabla}\alpha \times \boldsymbol{\nabla}\beta$, and $\delta\boldsymbol{A} = \boldsymbol{\nabla}\delta\gamma + \delta\alpha\boldsymbol{\nabla}\beta + \alpha\boldsymbol{\nabla}\delta\beta$. Therefore

$$\begin{aligned}\boldsymbol{B}\cdot\delta\boldsymbol{A} &= \boldsymbol{B}\cdot\boldsymbol{\nabla}\delta\gamma + \boldsymbol{B}\cdot\alpha\boldsymbol{\nabla}\delta\beta + (\boldsymbol{B}\cdot\boldsymbol{\nabla}\beta)\delta\alpha \\ &= \boldsymbol{\nabla}\cdot(\boldsymbol{B}\delta\gamma) + \boldsymbol{\nabla}\cdot(\boldsymbol{B}\alpha\delta\beta) - (\boldsymbol{B}\cdot\boldsymbol{\nabla}\alpha)\delta\beta + (\boldsymbol{B}\cdot\boldsymbol{\nabla}\beta)\delta\alpha.\end{aligned}$$

The last two terms vanish, so $\int d^3r\, \boldsymbol{A}\cdot\boldsymbol{B}$ is the surface term $\int d\boldsymbol{S}\cdot\boldsymbol{B}(\delta\gamma + \alpha\delta\beta)$ with no contribution from the bulk (finite $\boldsymbol{r}$ space). This of course is consistent with the Chern–Simons integral being a surface term, since $\delta\frac{1}{2}\int d^3r\, \boldsymbol{A}\cdot\boldsymbol{B} = \int d^3r\, \boldsymbol{B}\cdot\delta\boldsymbol{A}$. When demanding the variation of $\frac{1}{2}\int d^3r\, B^2$ to vanish, we first find $\int d^3r(\boldsymbol{\nabla}\times\boldsymbol{B})\cdot\delta\boldsymbol{A} = 0$. When $\delta\boldsymbol{A}$ is arbitrary, this condition implies the vanishing of $\boldsymbol{\nabla}\times\boldsymbol{B}$. However, in the Clebsch parameterization, all we can conclude is that $\boldsymbol{\nabla}\times\boldsymbol{B}$ is proportional to $\boldsymbol{B}$, with a position-dependent proportionality factor $\boldsymbol{\nabla}\times\boldsymbol{B} = \mu\boldsymbol{B}$. Taking the divergence shows that $\boldsymbol{B}\cdot\boldsymbol{\nabla}\mu = 0$, that is, μ can be a function of the magnetic surfaces; see (2.64), (2.69).

Problem 3.1. $\theta(t,\boldsymbol{r}) = r^2/2t$.
Time rescaling: $\theta_\omega(t,\boldsymbol{r}) = e^\omega\theta(T,\boldsymbol{r})$, $T = e^\omega t$;

$$e^\omega\theta(T,\boldsymbol{r}) = e^\omega r^2/2e^\omega t = r^2/2t = \theta(t,\boldsymbol{r}).$$

Space–time mixing: $\theta_{\boldsymbol{\omega}}(t,\boldsymbol{r}) = \theta(T,\boldsymbol{R})$,

$$\begin{aligned}&T = t + \boldsymbol{\omega}\cdot\boldsymbol{r} + \tfrac{1}{2}\omega^2\theta(T,\boldsymbol{R}) = t + \boldsymbol{\omega}\cdot\boldsymbol{r} + \omega^2R^2/4T,\\ &\boldsymbol{R} = \boldsymbol{r} + \boldsymbol{\omega}\theta(T,\boldsymbol{R}) = \boldsymbol{r} + \boldsymbol{\omega}R^2/2T.\end{aligned}$$

Squaring the second equation gives

$$R^2 = r^2 + \boldsymbol{\omega}\cdot\boldsymbol{r}R^2/T + \omega^2R^4/4T^2.$$

Multiplying the first equation by R^2/T gives

$$R^2 = tR^2/T + \boldsymbol{\omega}\cdot\boldsymbol{r}R^2/T + \omega^2R^4/4T^2.$$

Comparing the two shows that $R^2/T = r^2/t$ or $\theta(T,\boldsymbol{R}) = \theta(t,\boldsymbol{r})$.

Problem 6.1. From (6.23) and (6.26) we learn that $t = F'' + G''$, $x = (p+s)F'' - F' + (p-s)G'' - G'$, where F is a function of $p + s = R_+$ and G is a function of $p - s = R_-$. Differentiating these equations with respect to t and x, it follows that

$$1 = F''' \dot{R}_+ + G''' \dot{R}_-,$$

$$0 = F''' \frac{\partial R_+}{\partial x} + G''' \frac{\partial R_-}{\partial x},$$

$$0 = R_+ F''' \dot{R}_+ + R_- G''' \dot{R}_-,$$

$$1 = R_+ F''' \frac{\partial R_+}{\partial x} + R_- G''' \frac{\partial R_-}{\partial x},$$

which, in turn, imply

$$\frac{\partial R_+}{\partial x} = \frac{1}{(R_+ - R_-)} F''',$$

$$\frac{\partial R_-}{\partial x} = -\frac{1}{(R_+ - R_-)} G''',$$

and

$$\dot{R}_\pm = -R_\mp \frac{\partial R_\pm}{\partial x}$$

[the Riemann equation (6.19) again].

On the other hand, the functions $F_\pm(t \pm f)$ describing the Chaplygin gas solution from the Nambu–Goto equation are related to $R_\pm$ by (6.55): $R_\pm = 2F'_\pm$. Hence

$$\frac{\partial R_\pm}{\partial x} = \pm 2F''_\pm \frac{\partial f}{\partial x} = \pm \frac{2F''_\pm}{(F'_+ - F'_-)} = \pm \frac{4F''_\pm}{(R_+ - R_-)},$$

where (6.49b) is used. It follows that

$$4F''_+(z) F'''\left(2F'_+(z)\right) = \frac{d}{dz}\left(2F''\left(2F'_+(z)\right)\right) = 1$$

and

$$4F''_-(z) G'''\left(2F'_-(z)\right) = \frac{d}{dz}\left(2G''\left(2F'_-(z)\right)\right) = 1,$$

or

$$2F''\big(F'_+(z)\big) = z \qquad \text{and} \qquad 2G''\big(2F'_-(z)\big) = z.$$

When $F_+(z) = \ln z/2k$, $2F'_+(z) = 1/zk$; with $F(z) = z/2k \ln z$, $2F''(z) = 1/kz$ and $2F''\big(2F'_+(z)\big) = z$; similarly for $F_-(z)$ and $G(z)$.

References

[1] L. Landau and E. Lifshitz, *Fluid Mechanics,* 2nd ed. (Pergamon Press, Oxford, UK, 1987).

[2] V. Arnold and B. Khesin, *Topological Methods in Hydrodynamics* (Springer-Verlag, Berlin, 1998).

[3] The Lagrangian description of a fluid, and its relation to an Eulerian description, is explained, for example, in R. Salmon, *Ann. Rev. Fluid Mech.* **20**, 225 (1988).

[4] In the fluid mechanical context, these brackets were put forward by L. Landau, *Zh. Èksper. Teoret. Fiz.* **11**, 592 (1941) [English translation: *J. Phys. USSR* **5**, 71 (1941)]; see also P.J. Morrison and J.M. Greene, *Phys. Rev. Lett.* **45**, 790 (1980); Erratum, **48**, 569 (1982).

[5] L. Faddeev and R. Jackiw, *Phys. Rev. Lett.* **60**, 1692 (1988). For a detailed exposition, see R. Jackiw in *Constraint Theory and Quantization Methods*, F. Colomo, L. Lusanna, and G. Marmo, eds. (World Scientific, Singapore, 1994); reprinted in R. Jackiw, *Diverse Topics in Theoretical and Mathematical Physics* (World Scientific, Singapore, 1995).

[6] C. Eckart, *Phys. Rev.* **54**, 920 (1938).

[7] For an introduction to the properties of the Abelian and non-Abelian Chern–Simons terms, see S. Deser, R. Jackiw, and S. Templeton, *Ann. Physics (NY)* **140**, 372 (1982); Erratum, **185**, 406 (1985). In fluid mechanics or in magnetohydrodynamics the Abelian Chern–Simons term is known as the fluid or magnetic helicity, and was introduced by

L. Woltier, *Proc. Nat. Acad. Sci.* **44**, 489 (1958); and further studied in M. Berger and G. Field, *J. Fluid Mech.* **147**, 133 (1984); H. Moffatt and A. Tsinober, *Ann. Rev. Fluid Mech.* **24**, 281 (1992).

[8] C.C. Lin, *International School of Physics E. Fermi (XXI)*, G. Careri, ed. (Academic Press, New York, 1963).

[9] A. Clebsch, *J. Reine Angew. Math.* **56**, 1 (1859).

[10] A constructive discussion of the Darboux theorem can be found it the second work of [5].

[11] H. Lamb, *Hydrodynamics* (Cambridge University Press, Cambridge, UK, 1932), p. 248.

[12] R. Jackiw and S.-Y. Pi, *Phys. Rev. D* **61**, 105015 (2000).

[13] S. Treiman, R. Jackiw, B. Zumino, and E. Witten, *Current Algebra and Anomalies* (Princeton University Press, Princeton, NJ/World Scientific, Singapore, 1985).

[14] S. Deser, R. Jackiw, and A.P. Polychronakos, *Phys. Lett. A* **279**, 151 (2001).

[15] S. Chaplygin, *Sci. Mem. Moscow Univ. Math. Phys.* **21**, 1 (1904). [Chaplygin was a colleague of fellow USSR Academician N. Luzin. Although accused by Stalinist authorities of succumbing excessively to foreign influences, both unaccountably managed to escape the fatal consequences of their alleged actions; see N. Krementsov, *Stalinist Science* (Princeton University Press, Princeton NJ, 1997).] The same model (3.3) was later put forward by H.-S. Tsien, *J. Aeron. Sci.* **6**, 399 (1939) and by T. von Karman, *J. Aeron. Sci.* **8**, 337 (1941).

[16] K. Stanyukovich, *Unsteady Motion of Continuous Media* (Pergamon Press, Oxford, UK, 1960), p. 128.

[17] M. Hassaïne and P. Horvathy, *Ann. Physics (NY)* **282**, 218 (2000); L. O'Raifeartaigh and V. Sreedhar, *Ann. Physics (NY)* **293**, 215 (2001).

[18] R. Jackiw, *Phys. Today* **25**(1), 23 (1972); U. Niederer, *Helv. Phys. Acta* **45**, 802 (1972); C.R. Hagen, *Phys. Rev. D* **5**, 377 (1972).

[19] These symmetry transformations were identified by D. Bazeia and R. Jackiw, *Ann. Physics (NY)* **270**, 246 (1998), on the basis of constants of motion that were found previously; see [20].

[20] M. Bordemann and J. Hoppe, *Phys. Lett. B* **317**, 315 (1993); A. Jevicki, *Phys. Rev. D* **57**, 5955 (1988). These authors consider only the planar, $d = 2$, case.

[21] D. Bazeia, *Phys. Rev. D* **59**, 085007 (1999).

[22] Analytic details of the implementation of the space–time mixing transformation on (3.30) are presented by Bazeia and Jackiw [19].

[23] N. Ogawa, *Phys. Rev. D* **62**, 085023 (2000), who also presents other solutions, as does Bazeia [21].

[24] R. Jackiw and A.P. Polychronakos, *Comm. Math. Phys.* **207**, 107 (1999).

[25] The additional transformation rules were derived in [24], on the basis of constants of motion identified previously in the $d = 2$ case; see [26].

[26] M. Bordemann and J. Hoppe, *Phys. Lett. B* **325**, 359 (1994).

[27] G. Gibbons, *Nuclear Phys. B* **514**, 603 (1998).

[28] See, for example, Landau and Lifshitz [1] or S. Weinberg, *Gravitation and Cosmology* (Wiley, New York, 1972).

[29] That the theory of a membrane [(d = 2)-brane] in three spatial dimensions is equivalent to planar fluid mechanics was known to J. Goldstone (unpublished) and worked out by his student Hoppe (sometimes in collaboration with Bordemann). The method described in Sections 4.1 and 4.2 was presented for $d = 2$ in J. Hoppe, *Phys. Lett. B* **329**, 10 (1994), while a version of the argument in Section 4.3, specialized to $d = 2$, is found in Bordemann and Hoppe [20]. Generalization to arbitrary d is given in R. Jackiw and A.P. Polychronakos, *Proc. Steklov Inst. Math.* **226**, 193 (1999) and [24].

[30] Hassaïne and Horvathy [17].

[31] R. Jackiw and A.P. Polychronakos, *Phys. Rev. D* **62**, 085019 (2000). Some of these results are described in unpublished papers by J. Hoppe, Karlsruhe preprints KA-THEP-6-93 and KA-THEP-9-93 and hep-th/9311059.

[32] B. de Wit, J. Hoppe, and H. Nicolai, *Nuclear Phys. B* **305**, [FS23] 525 (1988).

[33] L. Susskind, *Phys. Rev.* **165**, 1535 (1968).

[34] Landau and Lifshitz [1]; Y. Nutku, *J. Math. Phys.* **28**, 2579 (1987); P. Olver and Y. Nutku, *J. Math. Phys.* **29**, 1610 (1988); M. Arik, F. Neyzi, Y. Nutku, P. Olver, and J. Verosky, *J. Math. Phys.* **30**, 1338 (1989); J. Brunelli and A. Das, *Phys. Lett. A* **235**, 597 (1997).

[35] B. Barbashov and N. Chernikov, *Zh. Èksper. Theoret. Fis.* **51**, 658 (1966) [English translation: *Soviet Phys. JETP* **24**, 437 (1967)].

[36] A. Jevicki and B. Sakita, *Phys. Rev. D* **22**, 467 (1980) and *Nuclear Phys. B* **165**, 511 (1980); J. Polchinski, *Nuclear Phys. B* **362**, 25 (1991); J. Avan and A. Jevicki, *Phys. Lett. B* **266**, 35 (1991) and **272**, 17 (1991).

[37] R. Floreanini and R. Jackiw, *Phys. Rev. Lett.* **59**, 1873 (1987).

[38] Y. Bergner and R. Jackiw, *Phys. Lett. A* **284**, 146 (2001).

[39] R. Jackiw, V.P. Nair, and S.-Y. Pi, *Phys. Rev. D* **62**, 085018 (2000).

[40] C. Crönstrom and J. Mickelsson, *J. Math. Phys.* **24**, 2528 (1983); Erratum, **27**, 419 (1986).

[41] H. Snyder, *Phys. Rev.* **71**, 38 (1947). This story was documented and told by J. Wess.

[42] G. Dunne, R. Jackiw, and C. Trugenberger, *Phys. Rev. D* **41**, 661 (1990); G. Dunne and R. Jackiw, *Nuclear Phys. C* (Proc. Suppl.) **33**, 114 (1993).

[43] R. Peierls, *Z. Phys.* **80**, 763 (1933).

[44] Z. Guralnik, R. Jackiw, S.-Y. Pi, and A.P. Polychronakos, *Phys. Lett. B* **517**, 450 (2001).

[45] For a review, see G. Dunne, *J. Phys. A* **21**, 2321 (1988).

[46] V. Arnold, *Mathematical Methods in Classical Mechanics*, (Springer-Verlag, New York, 1978).

[47] D.B. Fairlie, P. Fletcher, and C.K. Zachos, *Phys. Lett. B* **218**, 203 (1989).

[48] E. Mädelung, *Z. Phys.* **40**, 322 (1926); see also E. Merzbacher, *Quantum Mechanics* 3rd ed. (Wiley, New York, 1998).

[49] R. Jackiw and S.-Y. Pi, *Phys. Rev. Lett.* **66**, 2682 (1991).

Index